HOW THE LAWS OF PHYSICS LIE

HOW THE LAWS
OF PHYSICS LIE

NANCY CARTWRIGHT

CLARENDON PRESS · OXFORD
OXFORD UNIVERSITY PRESS · NEW YORK
1983

Oxford University Press, Walton Street, Oxford OX2 6DP

London Glasgow New York Toronto
Delhi Bombay Calcutta Madras Karachi
Kuala Lumpur Singapore Hong Kong Tokyo
Nairobi Dar es Salaam Cape Town
Melbourne Auckland
and associated companies in
Beirut Berlin Ibadan Mexico City Nicosia

OXFORD is a trade mark of Oxford University Press

Published in the United States
by Oxford University Press, New York

British Library Cataloguing in Publication Data

Cartwright, Nancy
How the laws of physics lie.
1. Physics—Philosophy
I. Title
530'.01   QC6
ISBN 0-19-824700-1
ISBN 0-19-824704-4 Pbk

Typeset by Joshua Associates (Oxford)
Printed in Great Britain
at the University Press, Oxford
by Eric Buckley
Printer to the University

To Marie and Claude

Acknowledgements

The author acknowledges permission to use previously published material in this volume. The provenance of the essays is as follows:

Essay 1. In part, from 'Causal Laws and Effective Strategies', *Noûs*, vol. 13 (1979). © *Noûs* 1979; reproduced by permission of *Noûs*.

In part, new.

Essay 2. From 'Truth Doesn't Explain Much', *American Philosophical Quarterly*, vol. 17 (1980).

Essay 3. In part, from 'Do the Laws of Physics State the Facts', *Pacific Philosophical Quarterly*, vol. 1 (1980).

In part, from 'How do We Apply Science', *PSA 1974 Proceedings*, ed. Robert Cohen *et al.* (Reidel, 1974).

In part, new.

Essay 4. From 'The Reality of Causes in a World of Instrumental Laws', *PSA 1980 Proceedings*, ed. P. Asquith and R. Giere (Philosophy of Science Association, 1980).

Essay 5. From 'When Explanation Leads to Inference', *Philosophical Topics*, special issue on Realism (forthcoming).

Essay 6. In part, from 'How Approximation Takes Us Away from Theory and Towards the Truth' by Nancy Cartwright and Jon J. Nordby, *Pacific Philosophical Quarterly* (forthcoming).

Essay 7. From 'Fitting Facts to Equations', *Philosophical Grounds of Rationality: Intentions, Categories, and Ends* (essays dedicated to Paul Grice), ed. Richard Grandy and Richard Warner (Oxford University Press, forthcoming).

Essay 8. New.

Essay 9. In part, from 'How the Measurement Problem is an Artefact of Mathematics', *Space, Time, and Causality*, ed. Richard Swinburne (Reidel, forthcoming).

In part, from *Studies in the Foundations of Quantum Mechanics*, ed. P. Suppes (Philosophy of Science Association, 1980).

In part, new.

Contents

Introduction   1

Essay 1   Causal Laws and Effective Strategies   21

Essay 2   The Truth Doesn't Explain Much   44

Essay 3   Do the Laws of Physics State the Facts?   54

Essay 4   The Reality of Causes in a World of
          Instrumental Laws   74

Essay 5   When Explanation Leads to Inference   87

Essay 6   For Phenomenological Laws   100

Essay 7   Fitting Facts to Equations   128

Essay 8   The Simulacrum Account of Explanation   143

Essay 9   How the Measurement Problem is an
          Artefact of the Mathematics   163

Author Index   217

Subject Index   219

Introduction

Philosophers distinguish phenomenological from theoretical laws. Phenomenological laws are about appearances; theoretical ones are about the reality behind the appearances. The distinction is rooted in epistemology. Phenomenological laws are about things which we can at least in principle observe directly, whereas theoretical laws can be known only by indirect inference. Normally for philosophers 'phenomenological' and 'theoretical' mark the distinction between the observable and the unobservable.

Physicists also use the terms 'theoretical' and 'phenomenological'. But their usage makes a different distinction. Physicists contrast 'phenomenological' with 'fundamental'. For example, Pergamon Press's *Encyclopaedic Dictionary of Physics* says, 'A phenomenological theory relates observed phenomena by postulating certain equations but does not enquire too deeply into their fundamental significance.'[1]

The dictionary mentions observed phenomena. But do not be misled. These phenomenological equations are not about direct observables that contrast with the theoretical entities of the philosopher. For look where this definition occurs—under the heading 'Superconductivity and superfluidity, phenomenological theories of'. Or notice the theoretical entities and processes mentioned in the contents of a book like *Phenomenology of Particles at High Energies* (proceedings of the 14th Scottish Universities Summer School in Physics): (1) Introduction to Hadronic Interactions at High Energies. (2) Topics in Particle Physics with Colliding Proton Beams. (3) Phenomenology of Inclusive Reactions. (4) Multihadron Production at High Energies: Phenomenology and Theory.[2]

---

[1] *The Encyclopaedic Dictionary of Physics* (Oxford: Pergamon Press, 1964), p. 108.
[2] P. L. Crawford and R. Jennings, *Phenomenology of Particles at High Energies* (London: Academic Press, 1974).

Francis Everitt, a distinguished experimental physicist and biographer of James Clerk Maxwell, picks Airy's law of Faraday's magneto-optical effect as a characteristic phenomenological law.[3] In a paper with Ian Hacking, he reports, 'Faraday had no mathematical theory of the effect, but in 1846 George Biddell Airy (1801–92), the English Astronomer Royal, pointed out that it could be represented analytically in the wave theory of light by adding to the wave equations, which contain second derivatives of the displacement with respect to time, other *ad hoc* terms, either first or third derivatives of the displacement.'[4] Everitt and Hacking contrast Airy's law with other levels of theoretical statement —'physical models based on mechanical hypotheses, . . . formal analysis within electromagnetic theory based on symmetry arguments', and finally, 'a physical explanation in terms of electron theory' given by Lorentz, which is 'essentially the theory we accept today'.

Everitt distinguishes Airy's phenomenological law from the later theoretical treatment of Lorentz, not because Lorentz employs the unobservable electron, but rather because the electron theory explains the magneto-optical effect and Airy's does not. Phenomenological laws describe what happens. They describe what happens in superfluids or meson-nucleon scattering as well as the more readily observed changes in Faraday's dense borosilicate glass, where magnetic fields rotate the plane of polarization of light. For the physicist, unlike the philosopher, the distinction between theoretical and phenomenological has nothing to do with what is observable and what is unobservable. Instead the terms separate laws which are fundamental and explanatory from those that merely describe.

The divide between theoretical and phenomenological commonly separates realists from anti-realists. I argue in these essays for a kind of anti-realism, and typically it is an anti-realism that accepts the phenomenological and rejects the theoretical. But it is not theory versus observation that I reject. Rather it is the theoretical as opposed to the phenomenological.

---

[3] In conversation, July 1981.

[4] C. W. F. Everitt and Ian Hacking, 'Theory or Experiment: Which Comes First?' *American Scientist*, forthcoming.

In modern physics, and I think in other exact sciences as well, phenomenological laws are meant to describe, and they often succeed reasonably well. But fundamental equations are meant to explain, and paradoxically enough the cost of explanatory power is descriptive adequacy. Really powerful explanatory laws of the sort found in theoretical physics do not state the truth.

I begin from the assumption that we have an immense number of very highly confirmed phenomenological laws. Spectra-physics Incorporated continuously runs a quarter of a million dollars' worth of lasers to death to test their performance characteristics. Nothing could be better confirmation than that. But how do the fundamental laws of quantum mechanics, which are supposed to explain the detailed behaviour of lasers, get their confirmation? Only indirectly, by their ability to give true accounts of lasers, or of benzene rings, or of electron diffraction patterns. I will argue that the accounts they give are generally not true, patently not true by the same practical standards that admit an indefinite number of commonplace phenomenological laws. We have detailed expertise for testing the claim of physics about what happens in concrete situations. When we look to the real implications of our fundamental laws, they do not meet these ordinary standards. Realists are inclined to believe that if theoretical laws are false and inaccurate, then phenomenological laws are more so. I urge just the reverse. When it comes to the test, fundamental laws are far worse off than the phenomenological laws they are supposed to explain.

The essays collected in this volume may be grouped around three different but interrelated arguments for this paradoxical conclusion.

(1) The manifest explanatory power of fundamental laws does not argue for their truth.

(2) In fact the way they are used in explanation argues for their falsehood. We explain by *ceteris paribus* laws, by composition of causes, and by approximations that improve on what the fundamental laws dictate. In all of these cases the fundamental laws patently do not get the facts right.

(3) The appearance of truth comes from a bad model of

explanation, a model that ties laws directly to reality. As an alternative to the conventional picture I propose a *simulacrum* account of explanation. The route from theory to reality is from theory to model, and then from model to phenomenological law. The phenomenological laws are indeed true of the objects in reality—or might be; but the fundamental laws are true only of objects in the model.

1.   AGAINST INFERENCE TO BEST EXPLANATION

I will argue that the falsehood of fundamental laws is a consequence of their great explanatory power. This is the exact opposite of what is assumed by a well-known and widely discussed argument form—inference to the best explanation. The basic idea of this argument is: if a hypothesis explains a sufficiently wide variety of phenomena well enough, we can infer that the hypothesis is true. Advocates of this argument form may disagree about what counts as well enough, or how much variety is necessary. But they all think that explanatory power, far from being at odds with truth, leads us to it. My first line of argument in these essays denies that explanation is a guide to truth.

Numerous traditional philosophical positions bar inferences to best explanations. Scepticism, idealism, and positivism are examples. But the most powerful argument I know is found in Pierre Duhem's *Aim and Structure of Physical Theory*,[5] reformulated in a particularly pointed way by Bas van Fraassen in his recent book *The Scientific Image*.[6] Van Fraassen asks, what has explanatory power to do with truth? He offers more a challenge than an argument: show exactly what about the explanatory relationship tends to guarantee that if x explains y and y is true, then x should be true as well. This challenge has an answer in the case of *causal* explanation, but *only* in the case of causal explanation. That is my thesis in 'When Explanation Leads to Inference'. Suppose we describe the concrete causal process by which a phenomenon is brought about. That kind of explanation

---

[5] Pierre Duhem, *The Aim and Structure of Physical Theory*, trans. Philip P. Wiener (New York: Atheneum, 1962).

[6] Bas van Fraassen, *The Scientific Image* (Oxford: Clarendon Press, 1980).

succeeds only if the process described actually occurs. To the extent that we find the causal explanation acceptable, we must believe in the causes described.

For example, consider the radiometer, invented by William Crookes in 1853. It is a little windmill whose vanes, black on one side, white on the other, are enclosed in an evacuated glass bowl. When light falls on the radiometer, the vanes rotate. At first it was assumed that light pressure causes the vanes to go round. Soon it was realized that the pressure of light would not be nearly great enough. It was then agreed that the rotation is due to the action of the gas molecules left inside the evacuated bowl. Crookes had tried to produce a vacuum in his radiometer. Obviously if we accept the agreed explanation, we infer that Crookes's vacuum was imperfect; the explanation demands the presence of molecules in the jar.

There were two rival hypotheses about what the molecules did. Both ideas are still defended by different camps today. A first proposal was that the vanes are pushed around by pressure of the molecules bouncing more energetically from the black side than the white. But in 1879 James Clerk Maxwell, using the kinetic theory of gases, argued that the forces in the gas would be the same in all directions, and so could not push the vanes. Instead differential heating in the gas produces tangential stresses, which cause slippage of the gas over the surface. As the gas flows around the edge, it pulls the vanes with it. In his biography of Maxwell, Francis Everitt urges the superiority of Maxwell's account over the more widely accepted alternative.[7] His confidence in Maxwell's causal story is reflected in his ontological views. His opponents think the tangential stresses are negligible. But unlike them, Everitt believes that if he builds a raidometer big enough he will be able to measure the flow of gas around the edge of the vanes.

The molecules in Crookes's radiometer are invisible, and the tangential stresses are not the kinds of things one would have expected to see in the first place. Yet, like Everitt, I believe in both. I believe in them because I accept Maxwell's

[7] C. W. F. Everitt, *James Clerk Maxwell—Physicist and Natural Philosopher* (New York: Charles Scribner's Sons, 1975). See Chapter 9.

causal account of why the vanes move around. In producing this account, Maxwell deploys certain fundamental laws, such as Boltzmann's equation and the equation of continuity, which I do not believe in. But one can reject theoretical laws without rejecting theoretical entities. In the case of Maxwell's molecules and the tangential stresses in the radiometer, there is an answer to van Fraassen's question: we have a satisfactory causal account, and so we have good reason to believe in the entities, processes, and properties in question.

Causal reasoning provides good grounds for our beliefs in theoretical entities. Given our general knowledge about what kinds of conditions and happenings are possible in the circumstances, we reason backwards from the detailed structure of the effects to exactly what characteristics the causes must have in order to bring them about. I have sometimes summarized my view about explanation this way: no inference to best explanation; only inference to most likely cause. But that is right only if we are very careful about what makes a cause 'likely'. We must have reason to think that this cause, and no other, is the only practical possibility, and it should take a good deal of critical experience to convince us of this.

We make our best causal inferences in very special situations—situations where our general view of the world makes us insist that a known phenomenon has a cause; where the cause we cite is the kind of thing that could bring about the effect and there is an appropriate process connecting the cause and the effect; and where the likelihood of other causes is ruled out. This is why controlled experiments are so important in finding out about entities and processes which we cannot observe. Seldom outside of the controlled conditions of an experiment are we in a situation where a cause can legitimately be inferred.

Again the radiometer illustrates. Maxwell is at odds with the standard account. To resolve the debate, Maxwell's defender, Everitt, proposes not further theoretical analysis, but rather an experiment. He wants to build an enormous radiometer, where he can control the partial vacuum and its viscosity, vary the coefficient of friction on the vanes, vary their widths, take into account winds in the jar, and finally determine whether the tangential stresses are really the major cause of rotation.

The dispute about normal and tangential stresses highlights a nice point about observation. Philosophical debate has focused on entities. Instrumentalists, who want to believe only in what they can see, get trapped in footling debates: do we really 'see' through a microscope? An electron microscope? A phase-interference light microscope? Even with the naked eye, do we not in any case see only effects? But many of the things that are realities for physics are not things to be seen. They are non-visual features—the spin of the electron, the stress between the gas surface, the rigidity of the rod. Observation—seeing with the naked eye—is not the test of existence here. Experiment is. Experiments are made to isolate true causes and to eliminate false starts. That is what is right about Mill's 'methods'.

Where can such an idea make a difference? I think these are just the kinds of considerations that need to be brought to current philosophical debates about quantum electrodynamics. Nobody denies the enormous organizing and predictive power of this theory, especially since the development in the last few years of gauge field theories, which unify weak and electromagnetic phenomena. Many think that quantum electrodynamics is the most impressive theory we have ever had in these respects. But as elementary particle physicist James Cushing remarks,

When one looks at the succession of blatantly *ad hoc* moves made in QFT [quantum field theory] (negative-energy sea of electrons, discarding of infinite self energies and vacuum polarizations, local gauge invariance, forcing renormalization in gauge theories, spontaneous symmetry breaking, permanently confined quarks, color, just as examples) and of the picture which emerges of the 'vacuum' (aether?), as seething with particle–antiparticle pairs of every description and as responsible for breaking symmetries initially present, one can ask whether or not nature is *seriously* supposed to be like that.[8]

Does the success of quantum field theory argue for the existence of negative energy electrons, permanently confined quarks, and a vacuum 'seething with particle–antiparticle

[8] James Cushing, 'Models and Methodologies in Current Theoretical High-Energy Physics', *Synthese* 50 (1982), p. 78.

pairs of every description'? Debate among philosophers has tended to focus on the coherence of the theory, or on the true extent of is successes.[9] I think we should instead focus on the causal roles which the theory gives to these strange objects: exactly how are they supposed to bring about the effects which are attributed to them, and exactly how good is our evidence that they do so? The general success of the theory at producing accurate predictions, or at unifying what before had been disparate, is no help here. We can believe in the unexpected entities of quantum electrodynamics if we can give them concrete causal roles; and the rationality of that belief will depend on what experimental evidence supports the exact details of those causal claims.

Although I claim that a successful causal explanation gives good reason to believe in the theoretical entities and theoretical properties it postulates, I have repeatedly said that I do not believe in theoretical laws. But do not properties and laws go hand-in-hand? Bas van Fraassen asks: are not 'inferences to causes after all merely inferences to the truth of propositions describing general characteristics of . . . the things the propositions are about'?[10] The answer to van Fraassen's question is, undoubtedly, yes. But the propositions to which we commit ourselves when we accept a causal explanation are highly detailed causal principles and concrete phenomenological laws, specific to the situation at hand, not the abstract equations of a fundamental theory. Maxwell says that the vanes are dragged around by gas sliding over the edge. They are not pushed by light pressure or the normal force of the gas on the surface. The acceptability of his account depends on a host of general claims about what happens in radiometers.

Here is one phenomenological law—in this case a causal principle—which Maxwell uses:

[The] velocity (with which the gas slides over the surface) and the corresponding tangential stress are affected by inequalities of temperature

---

[9] See the January 1982 issue of *Synthese*, which is devoted to this topic, and further references therein.

[10] In correspondence in June 1981.

at the surface of the solid, which give rise to a force tending to make the gas slide along the surface from colder to hotter places.[11]

Here is another, this one critical to his argument that the vanes are not pushed by pressure normal to the surface:

When the flow of heat is steady, these forces (the total forces acting in all directions) are in equilibrium.[12]

Maxwell's explanation of exactly how the motion in radiometers takes place will not be right unless these principles are true. But these are not fundamental laws. Maxwell sets his particular causal story into the framework of the developing kinetic theory of gases. It is useful to contrast the two specific laws quoted, about what happens in radiometers, with two fundamental equations from this basic theory which Maxwell uses. In his derivation, he employs both Boltzmann's equation

$$\frac{df_1}{dt} + \xi_1 \frac{df_1}{dx} + \eta_1 \frac{df_1}{dy} + \zeta_1 \frac{df_1}{dz} + X\frac{df_1}{d\xi_1} + Y\frac{df_1}{d\eta_1} + Z\frac{df_1}{d\zeta_1} +$$

$$+ \iiint d\xi_2 d\eta_2 d\zeta_2 \int b\,db \int d\phi V(f_1 f_2 - f_1' f_2') = 0, \tag{1}$$

and the general equation of continuity

$$\frac{\partial}{\partial t}[Q\rho] + \frac{d}{dx}[Q(u + \xi - U)] + \frac{d}{dy}[Q(v + \eta - V] +$$

$$+ \frac{d}{dz}[Q(w + \zeta - W)] = \rho \frac{\delta}{\delta t} Q. \tag{2}$$

These are general, abstract equations; they are not about any particular happenings in any particular circumstances. The contrast is like that between moral principles, as Aristotle sees them in the *Nicomachean Ethics*, Book II, Chapter 7, 'Among statements about conduct those which are general apply more widely, but those which are particular are more genuine.'

---

[11] James Clerk Maxwell, 'On Stresses in Rarified Gases Arising from Inequalities of Temperature', *The Scientific Papers of James Clerk Maxwell* ed. W. D. Niven (New York: Dover Publishers, 1965), p. 703.
[12] Ibid., p. 684.

Explanatory power is no guarantee of truth, unless van Fraassen's challenge can be met. I argue that, in the very special case of causal explanation, the challenge is met. In causal explanations truth is essential to explanatory success. But it is only the truth of low-level causal principles and concrete phenomenological laws. Is there no further account that secures the truth of abstract laws as well; no story of explanation that shows that abstract laws must be true if they are to explain? There are two models of theoretical explanation that could do so. I discuss them in Essay 6. Both have serious flaws. The other essays in this volume that argue against inference to the best explanation are 'When Explanation Leads to Inference', and 'The Reality of Causes in a World of Instrumental Laws'.

Since I make such heavy use of the notion of causal principles—a notion which empiricists will be wary of—one earlier paper is included. 'Causal Laws and Effective Strategies' argues that causal laws are quite as objective as the more Humean laws of association. One standard point of view maintains that causal laws are required to account for explanatory asymmetries. If only laws of association are admitted, the length of the shadow can as well explain the height of the flagpole as the reverse. In *The Scientific Image* van Fraassen argues persuasively that these asymmetries are not genuine. I think he is mistaken. But his case is powerful, and it might persuade us to give up certain explanatory strategies. But we will not so easily be persuaded to give up our strategies for action, which are essential to practical life. Perhaps there is no fact of the matter about what explains what. But there is no doubt that spraying swamps is an effective way to stop the spread of malaria, whereas burning the blankets of malarial patients is not. The essay on causation argues that laws of association are insufficient to account for the facts about effective strategies. Causal laws are required as well. Besides defending this central philosophical thesis the first essay re-introduces Simpson's paradox to the philosophical literature and it finds in Simpson's paradox the underlying source of a variety of different counterexamples that philosophers have proposed against probabilistic models of causation.

1.1. *Composition of causes*

Explaining in physics involves two quite different kinds of activities. First, when we explain a phenomenon, we state its causes. We try to provide detailed accounts of exactly how the phenomenon is produced. Second, we fit the phenomenon into a broad theoretical framework which brings together, under one set of fundamental equations, a wide array of different kinds of phenomena. Both kinds of explanations use what philosophers have called laws of nature, but as we have seen in the case of the radiometer, the laws for the two kinds of explanation do not look at all alike. The causal story uses highly specific phenomenological laws which tell what happens in concrete situations. But the theoretical laws, like the equation of continuity and Boltzmann's equation, are thoroughly abstract formulae which describe no particular circumstances.

The standard covering-law account tries to fit both kinds of explanation into the same mould. But the function of the laws is different in the two cases, and so too, I have argued, are their claims to truth. The difference is more than philosophical. We find it in scientific practice (cf. 'The Reality of Causes in a World of Instrumental Laws'). In physics it is usual to give alternative theoretical treatments of the same phenomenon. We construct different models for different purposes, with different equations to describe them. Which is the right model, which the 'true' set of equations? The question is a mistake. One model brings out some aspects of the phenomenon; a different model brings out others. Some equations give a rougher estimate for a quantity of interest, but are easier to solve. No single model serves all purposes best.

Causal explanation is different. We do not tell first one causal story then another, according to our convenience. Maxwell's explanation involving tangential stresses in the radiometer is incompatible with the earlier light pressure account, and it is incompatible with the more standard hypothesis involving normal pressures. If one of these is adopted, the others are rejected. Alternative causal stories compete in physics in a way in which theoretical treatments do not. Causal stories are treated as if they are true or false, but which theoretical laws 'govern' the phenomenon is a matter of convenience.

Perhaps laws in physics are not deployed in explanations as if they are true. But I claim something stronger: if the evidence is taken seriously, they must be judged false. Why do I urge this far stronger claim? One reason is the tension between causal explanation and theoretical explanation. Physics aims to give both, but the needs of the two are at odds with one another. One of the important tasks of a causal explanation is to show how various causes combine to produce the phenomenon under study. Theoretical laws are essential in calculating just what each cause contributes. But they cannot do this if they are literally true; for they must ignore the action of laws from other theories to do the job.

The third essay in this volume asks, 'Do the Laws of Physics State the Facts?' I answer *no*. When different kinds of causes compose, we want to explain what happens in the intersection of different domains. But the laws we use are designed only to tell truly what happens in each domain separately.[13] This is also the main theme of the second essay.

Realists are inclined to invoke the unity of nature in reply: the true explanation in cases where causes combine comes from a 'super' law which unifies the separate domains. I am dubious about the existence of these unifying laws. Later I will also claim that I do not believe there are enough bridge laws of a certain sort. My reasons in both cases are the same. I think we should believe only in laws for which we have evidence. Maxwell *showed* that electromagnetism and light could be treated together under the same theoretical umbrella by producing Maxwell's theory, which gives marvellously successful accounts of both. The brilliant applied mathematician and cosmologist, Stephen Hawking, entitled his inaugural lecture for the Plumean Professorship at Cambridge, 'Is the End in Sight for Theoretical Physics?' He has an immense and excited confidence that he and his colleagues

---

[13] There is a very nice paper by Geoffrey Joseph, 'The Many Sciences and the One World', *Journal of Philosophy* 77 (1980), pp. 773–90, that argues this same point.

are on the verge of writing down the right equations to unify the basic forces in nature. We should agree that the end of theoretical physics is in view only when it is clear they have done so.

A second reason why I do not believe in these unified laws is methodological. It runs throughout the essays collected here. In metaphysics we try to give general models of nature. We portray it as simple or complex, law-governed or chancy, unified or diverse. What grounds do we have for our choices? *A priori* intuitions and abstract arguments are not good enough. We best see what nature is like when we look at our knowledge of it. If our best-supported theories now are probabilistic, we should not insist on determinism. If Russell was right that physics does not employ causes, we should agree with Hume, at least about the basic material phenomena studied by physics. Unity of science is a case in point. How unified is our knowledge? Look at any catalogue for a science or engineering school. The curriculum is divided into tiny, separate subjects that irk the interdisciplinist. Our knowledge of nature, nature as we best see it, is highly compartmentalized. Why think nature itself is unified?

So far I have concentrated on the composition of causes. But the problems raised by the composition of causes are just a special case. Even if we do not cross domains or study causes which fall under different basic laws, still the use of fundamental laws argues for their falsehood. If the fundamental laws are true, they should give a correct account of what happens when they are applied in specific circumstances. But they do not. If we follow out their consequences, we generally find that the fundamental laws go wrong; they are put right by the judicious corrections of the applied physicist or the research engineer.

'For Phenomenological Laws' argues this point. Much of this essay is taken from a joint paper with Jon Nordby[14] and it begins from a view we both share: there are no rigorous solutions for real life problems. Approximations and adjustments are required whenever theory treats reality. As an

[14] Jon Nordby and Nancy Cartwright, 'How Approximation Takes Us Away from Theory and Towards the Truth' (Pacific Lutheran University and Stanford University: unpublished manuscript).

example of the extent of the shortfall let us look at the introduction to an advanced text, *Perturbation Methods in Fluid Mechanics* by Milton Van Dyke.

> Because of this basic non-linearity, exact solutions are rare in any branch of fluid mechanics . . . So great is the need that a solution is loosely termed 'exact' even when an ordinary differential equation must be integrated numerically. Lighthill (1948) has given a more or less exhaustive list of such solutions for inviscid compressible flow:[15]

Van Dyke lists seven cases, then continues,

> Again, from Schlichting (1968) one can construct a partial list for incompressible viscous flow:[16]

The second list gives seven more examples. But even these fourteen examples do not provide a rigorous tie between fundamental theory and practical circumstance. Van Dyke concludes:

> It is typical of these self-similar flows that they involve idealized geometries far from most shapes of practical interest. To proceed further one must usually approximate.[17]

Realistically-oriented philosophers are inclined to think that approximations raise no problems in principle. The 'true' solution is the rigorous solution, and departures from it are required only because the mathematics is too difficult or too cumbersome. Nordby calls approximations that really aim to estimate the rigorous results *ab vero* approximations.[18] *Ab vero* approximations seem to suit the realist's case well, but 'For Phenomenological Laws' argues that even these do not provide positive evidence for the truth of fundamental laws. Worse for the realist is the widespread use of *ad verum* approximation. Here the approximation goes in the opposite direction. The steps in the derivation move away from the rigorous consequences of the starting laws, correcting and improving them, in order to arrive finally

---

[15] Milton Van Dyke, *Perturbation Methods in Fluid Mechanics* (Standford: Parabolic Press, 1975), p. 1.
[16] Ibid., p. 1.
[17] Ibid., p. 2.
[18] Jon Nordby, 'Two Kinds of Approximation in the Application of Science' (Pacific Lutheran University: unpublished manuscript).

at an accurate description of the phenomena. Two kinds of illustrations are given in 'For Phenomenological Laws', both taken from the joint paper with Nordby. Both undermine the realist's use of inference to the best explanation: the application of laws to reality by a series of *ad verum* approximations argues for their falsehood, not their truth.

Approximations enter when we go from theory to practice. Consider the reverse direction, not 'theory exit' but 'theory entry'. In theory entry, we begin with a factual description, and look to see how it can be brought under a fundamental law or equation. The canonical method is via a bridge principle. But that proposal rests on a too-simple view of how explanations work. To get from a detailed factual knowledge of a situation to an equation, we must prepare the description of the situation to meet the mathematical needs of the theory. Generally the result will no longer be a true description. In 'Fitting Facts to Equations' I give some simple examples of the kinds of descriptions for which we have equations in quantum mechanics. Look there to see how different these are from the kinds of descriptions we would give if we wanted an accurate report of the facts.

Contrary to the conventional account, which relies solely on bridge principles, I think theory entry proceeds in two stages. We start with an *unprepared* description which gives as accurate a report as possible of the situation. The first stage converts this into a *prepared* description. At the second stage the prepared description is matched to a mathematical representation from the theory. Ideally the prepared description should be true to the unprepared. But the two activities pull in opposite directions, and a description that is adequate to the facts will seldom have the right mathematical structure. All this is argued in Essay 7. What I call there 'preparing a description' is exactly what we do when we produce a model for a phenomenon, and the two-stage view of theory entry in that paper lays the groundwork for the account in Essay 8 which places models at the core of explanation.

1.2 *An alternative to the covering-law model of explanation*

The third line of argument offers an alternative to the covering-law model of explanation. Although I do think that we can give causal explanations of isolated events, I shall discuss here only explanations for kinds of events that recur in a regular way, events that can be described by phenomenological laws. I am thinking of the kinds of explanation that are offered in highly mathematical theories.

I said in the last section that there are two quite different kinds of things we do when we explain a phenomenon in physics. First, we describe its causes. Second, we fit the phenomenon into a theoretical frame. Ever since the earliest expositions of the covering-law model, it has been objected that its account of the first kind of explanation is inadequate. I have been strongly influenced by the criticism due to Michael Scriven[19] and Alan Donagan,[20] but many others make similar points as well. At present Wesley Salmon[21] is developing an alternative account of this kind of explanation, which focuses on singular causal processes and on causal interactions. Here I consider only the second kind of explanatory activity. How do we fit a phenomenon into a general theoretical framework?

Prima facie, the covering-law model seems ideally suited to answer: we fit a phenomenon into a theory by showing how various phenomenological laws which are true of it derive from the theory's basic laws and equations. This way of speaking already differentiates me from the covering-law theorist. I do not talk about explaining a feature of a phenomenon by deriving a description of that feature; but rather of treating a phenomenon by deriving a variety of phenomenological laws about it. But this is not the primary difference. The 'covering' of 'covering-law model' is a powerful metaphor. It teaches not only that phenomenological laws can be derived from fundamental laws, but also that the fundamental laws are laws that govern the phenomena. They are laws that cover the phenomena, perhaps under a more general or abstract description, perhaps in virtue of some

[19] See Michael Scriven, 'Causes Connections and Conditions in History' in William H. Dray (ed.) *Philosophical Analysis and History* (New York: Harper & Row, 1966).

[20] See Alan Donagan, 'Explanations in History' in P. Gardiner (ed.), *Theories of History* (Glencoe, Illinois: The Free Press, 1959).

[21] See Wesley Salmon's new book tentatively titled *Scientific Explanation and A Causal Structure of the World* (Princeton: Princeton University Press, forthcoming).

hidden micro-structural features; but still the fundamental laws apply to the phenomena and describe how they occur.

I propose instead a 'simulacrum' account. That is not a word we use any more, but one of its dictionary definitions captures exactly what I mean. According to the second entry in the Oxford English Dictionary, a simulacrum is 'something having merely the form or appearance of a certain thing, without possessing its substance or proper qualities'. On the simulacrum account, to explain a phenomenon is to construct a model which fits the phenomenon into a theory. The fundamental laws of the theory are true of the objects in the model, and they are used to derive a specific account of how these objects behave. But the objects of the model have only 'the form or appearance of things' and, in a very strong sense, not their 'substance or proper qualities'.

The covering-law account supposes that there is, in principle, one 'right' explanation for each phenomenon. The simulacrum account denies this. The success of an explanatory model depends on how well the derived laws approximate the phenomenological laws and the specific causal principles which are true of the objects modelled. There are always more phenomenological laws to be had, and they can be approximated in better and in different ways. There is no single explanation which is the right one, even in the limit, or relative to the information at hand. Theoretical explanation is, by its very nature, redundant. This is one of the endemic features of explanation in physics which the deductive-nomological (D-N) account misses, albeit with the plea that this annoying feature will no longer be present when the end of physics is achieved.

'Fitting Facts to Equations' is full of examples of the kinds of models we use in explaining things in physics. I offer the simulacrum account in the next essay as an alternative which provides a better description of actual explanatory practice than do conventional covering-law accounts. It obviously serves my attack on fundamental laws. The 'simulacrum' aspect is intended seriously: generally the prepared and unprepared descriptions cannot be made to match. But only the prepared descriptions fall under the basic laws. The lesson for the truth of fundamental laws is clear: fundamental laws do not govern objects in reality; they govern only objects in models.

2.   WHAT DIFFERENCE DOES IT MAKE?

Debate between realists and non-realists has been going on for a long time. Does the outcome have any practical consequence? I think it does. The last essay in this volume provides one example. I have argued that many abstract concepts in physics play merely an organizing role and do not seem to represent genuine properties. Unitarity has the earmarks of being just such a concept in quantum mechanics.

Unitarity is a property of operators. Those operators that are unitary represent motions that are indeterministic. Unitarity plays no causal role in the theory; nothing else about its use argues for interpreting it as a real property either. Yet there is a tendency to think of it not only as a mathematical characteristic of operators, but also as a genuine property of the situations represented by the operators. This, I claim, is the source of the notorious measurement problem in quantum mechanics. Unitarity marks no real property in quantum theory, and if we do not suppose that it must do so we have no interesting philosophical problem about measurement.

Essay 9 is concerned exclusively with the measurement problem, and not with the other conceptual difficulty in quantum mechanics that philosophers commonly discuss, the Einstein–Podolsky–Rosen paradox. These two problems tend to sit at opposite ends of a balance: philosophical treatments that offer hope for solving one usually fare badly with the other. This is certainly the case with the programme I propose. If it should work, it would at best eliminate the measurement problem, which I believe to be a pseudo-problem. But it would have nothing to say regarding the very perplexing facts about locality brought out by the EPR paradox.

3.   CONCLUSION

The picture of science that I present in these essays lacks the purity of positivism. It is a jumble of unobservable entities, causal processes, and phenomenological laws. But it shares one deep positivist conviction: there is no better reality besides the reality we have to hand. In the second sentence of this introduction I characterized the distinction between phenomenological and theoretical laws: phenomenological laws are about appearances; theoretical ones about the reality behind the appearances. That is the distinction I reject. Richard Feynmann talks about explaining in physics as fitting phenomena into 'the patterns of nature'. But where are the patterns? Things happen in nature. Often they happen in regular ways when the circumstances are similar; the same kinds of causal processes recur; there are analogies between what happens in some situations and what happens in others. As Duhem suggests, what happens may even be organized into natural kinds in a way that makes prediction easy for us (see the last section of 'When Explanation Leads to Inference'). But there is only what happens, and what we say about it. Nature tends to a wild profusion, which our thinking does not wholly confine.

The metaphysical picture that underlies these essays is an Aristotelean belief in the richness and variety of the concrete and particular. Things are made to look the same only when we fail to examine them too closely. Pierre Duhem distinguished two kinds of thinkers: the deep but narrow minds of the French, and the broad but shallow minds of the English. The French mind sees things in an elegant, unified way. It takes Newton's three laws of motion and turns them into the beautiful, abstract mathematics of Lagrangian mechanics. The English mind, says Duhem, is an exact contrast. It engineers bits of gears, and pulleys, and keeps the strings from tangling up. It holds a thousand different details all at once, without imposing much abstract order or organization. The difference between the realist and me is almost theological. The realist thinks that the creator of the universe worked like a French mathematician. But I think that God has the untidy mind of the English.

GUIDE FOR THE READER

The last essay on quantum mechanics shows how anti-realism can be put to work. The first essay defends causes. The principal arguments for theoretical entities and against theoretical laws are in the middle essays. Although the essays argue in favour of theoretical entities and against theoretical laws, the main emphasis is on the latter theme. This book is a complement, I think, to the fine discussions of representation, experimentation, and creation of phenomena in Ian Hacking's *Representing and Intervening*.[22] Hacking provides a wealth of examples which show how new entities are admitted to physics. Essay 6 here has some detailed examples and many equations that may not be of interest to the reader with pure philosophic concerns; but the general point of the examples can be gleaned by reading the introductory sections. Although Essay 9 is about quantum mechanics, it is not technical and readers without expert knowledge will be able to follow the argument.

[22] Ian Hacking, *Representing and Intervening* (Cambridge: Cambridge University Press, forthcoming).

Essay 1

Causal Laws and Effective Strategies

0.   INTRODUCTION

There are at least two kinds of laws of nature: laws of association and causal laws. Laws of association are the familiar
laws with which philosophers usually deal. These laws tell
how often two qualities or quantities are co-associated. They
may be either deterministic—the association is universal—or
probabilistic. The equations of physics are a good example:
whenever the force on a classical particle of mass m is f the
acceleration is f/m. Laws of association may be time indexed,
as in the probabilistic laws of Mendelian genetics, but,
apart from the asymmetries imposed by time indexing,
these laws are causally neutral. They tell how often two
qualities co-occur; but they provide no account of what
makes things happen.

Causal laws, by contrast, have the word 'cause'—or
some causal surrogate—right in them. Smoking causes lung
cancer; perspiration attracts wood ticks; or, for an example
from physics, force causes change in motion: to quote
Einstein and Infeld, 'The action of an external force changes
the velocity . . . such a force either increases or decreases
the velocity according to whether it acts in the direction
of motion or in the opposite direction.'[1]

Bertrand Russell argued that laws of association are all
the laws there are, and that causal principles cannot be
derived from the causally symmetric laws of association.[2]
I shall here argue in support of Russell's second claim, but
against the first. Causal principles cannot be reduced to
laws of association; but they cannot be done away with.

The argument in support of causal laws relies on some
facts about strategies. They are illustrated in a letter which

[1] Albert Einstein and Leopold Infeld, *The Evolution of Physics* (Cambridge:
Cambridge University Press, 1971), p. 9.
[2] Bertrand Russell, 'On the Notion of Cause', *Proceedings of the Aristotelian
Society* 13 (1912–13), pp. 1–26.

I recently received from TIAA–CREF, a company that provides insurance for college teachers. The letter begins:

It simply wouldn't be true to say,
    'Nancy L. D. Cartwright . . . if you own a TIAA life insurance policy you'll live longer.'
But it is a fact, nonetheless, that persons insured by TIAA do enjoy longer lifetimes, on the average, than persons insured by commercial insurance companies that serve the general public.

I will take as a starting point for my argument facts like those reported by the TIAA letter: *it wouldn't be true that* buying a TIAA policy would be an effective strategy for lengthening one's life. TIAA may, of course, be mistaken; it could after all be true. What is important is that their claim is, as they suppose, the kind of claim which is either true or false. There is a pre-utility sense of goodness of strategy; and what is and what is not a good strategy in this pre-utility sense is an objective fact. Consider a second example. Building the canal in Nicaragua, the French discovered that spraying oil on the swamps is a good strategy for stopping the spread of malaria, whereas burying contaminated blankets is useless. What they discovered was true, independent of their theories, of their desire to control malaria, or of the cost of doing so.

The reason for beginning with some uncontroversial examples of effective and ineffective strategies is this: I claim causal laws cannot be done away with, for they are needed to ground the distinction between effective strategies and ineffective ones. If indeed, it *isn't true that* buying a TIAA policy is an effective way to lengthen one's life, but stopping smoking is, the difference between the two depends on the causal laws of our universe, and on nothing weaker. This will be argued in Part 2. Part 1 endorses the first of Russell's claims, that causal laws cannot be reduced to laws of association.

1.   STATISTICAL ANALYSES OF CAUSATION

I will abbreviate the causal law, 'C causes E' by $C \hookrightarrow E$. Notice that C and E are to be filled in by general terms, and not names of particulars; for example, 'Force causes motion' or 'Aspirin relieves headache'. The generic law 'C causes E' is not to be understood as a universally quantified law about particulars, even about particular causal facts. It is generically true that aspirin relieves headache even though some particular aspirins fail to do so. I will try to explain what causal laws assert by giving an account of how causal laws relate on the one hand to statistical laws, and on the other to generic truths about strategies. The first task is not straightforward; although causal laws are intimately connected with statistical laws, they cannot be reduced to them.

A primary reason for believing that causal laws cannot be reduced to probabilistic laws is broadly inductive: no attempts so far have been successful. The most notable attempts recently are by the philosophers Patrick Suppes[3] and Wesley Salmon[4] and, in the social sciences, by a group of sociologists and econometricians working on causal models, of whom Herbert Simon and Hubert Blalock[5] are good examples.

It is not just that these attempts fail, but rather why they fail that is significant. The reason is this. As Suppes urges, a cause ought to increase the frequency of its effect. But this fact may not show up in the probabilities if other causes are at work. Background correlations between the purported cause and other causal factors may conceal the increase in probability which would otherwise appear. A simple example will illustrate.

It is generally supposed that smoking causes heart disease ($S \hookrightarrow H$). Thus we may expect that the probability of heart disease on smoking is greater than otherwise. (We can write this as either $\text{Prob}(H/S) > \text{Prob}(H)$, or $\text{Prob}(H/S) > \text{Prob}(H/\neg S)$, for the two are equivalent.) This expectation is mistaken. Even if it is true that smoking causes heart disease, the expected increase in probability will not appear if smoking is correlated with a sufficiently strong preventative, say

[3] See Patrick Suppes, *A Probabilistic Theory of Causality* (Amsterdam: North-Holland Publishing Co., 1970).

[4] See Wesley Salmon, 'Statistical Explanation' in Salmon, Wesley (ed.), *Statistical Explanation and Statistical Relevance* (Pittsburgh: University of Pittsburgh Press, 1971).

[5] See H. M. Blalock, Jr., *Causal Models in the Social Sciences* (Chicago: Aldine-Atherton, 1971).

exercising. (Leaving aside some niceties, we can render 'Exercise prevents heart disease' as $X \hookrightarrow \neg H$.) To see why this is so, imagine that exercising is more effective at preventing heart disease than smoking at causing it. Then in any population where smoking and exercising are highly enough correlated,[6] it can be true that $\text{Prob}(H/S) = \text{Prob}(H)$, or even $\text{Prob}(H/S) < \text{Prob}(H)$. For the population of smokers also contains a good many exercisers, and when the two are in combination, the exercising tends to dominate.

It is possible to get the increase in conditional probability to reappear. The decrease arises from looking at probabilities that average over both exercisers and non-exercisers. Even though in the general population it seems better to smoke than not, in the population consisting entirely of exercisers, it is worse to smoke. This is also true in the population of non-exercisers. The expected increase in probability occurs not in the general population but in both sub-populations.

This example depends on a fact about probabilities known as Simpson's paradox,[7] or sometimes as the Cohen–Nagel–Simpson paradox, because it is presented as an exercise in Morris Cohen's and Ernest Nagel's text, *An Introduction to Logic and Scientific Method*.[8] Nagel suspects that he learned about it from G. Yule's *An Introduction to the Theory of Statistics* (1904), which is one of the earliest textbooks written on statistics; and indeed it is dicussed at length there. The fact is this: any association—$\text{Prob}(A/B) = \text{Prob}(A)$; $\text{Prob}(A/B) > \text{Prob}(A)$; $\text{Prob}(A/B) < \text{Prob}(A)$— between two variables which holds in a given population can be reversed in the sub-populations by finding a third variable which is correlated with both.

In the smoking–heart disease example, the third factor is a preventative factor for the effect in question. This is just one possibility. Wesley Salmon[9] has proposed different examples to show that a cause need not increase the probability of its effect. His examples also turn on Simpson's

---

[6] Throughout, '*A* and *B* are correlated' will mean $\text{Prob}(A/B) \neq \text{Prob}(A)$.

[7] E. H. Simpson, 'The Interpretation of Interaction in Contingency Tables', *Journal of the Royal Statistical Society*, Ser. B. 13 (1951), pp. 238–41.

[8] See Morris R. Cohen and Ernest Nagel, *An Introduction to Logic and Scientific Method* (New York: Harcourt, Brace and Co., 1934).

[9] See Wesley Salmon, op. cit.,

paradox, except that in his cases the cause is correlated, not with the presence of a negative factor, but with the absence of an even more positive one.

Salmon considers two pieces of radioactive material, uranium 238 and polonium 214. We are to draw at random one material or the other, and place it in front of a Geiger counter for some time. The polonium has a short half-life, so that the probability for some designated large number of clicks is .9; for the long-lived uranium, the probability is .1. In the situation described, where one of the two pieces is drawn at random, the total probability for a large number of clicks is $\frac{1}{2}(.9) + \frac{1}{2}(.1) = .5$. So the conditional probability for the Geiger counter to click when the uranium is present is less than the unconditional probability. But when the uranium has been drawn and the Geiger counter does register a large number of clicks, it is the uranium that causes them. The uranium decreases the probability of its effect in this case. But this is only because the even more effective polonium is absent whenever the uranium is present.

All the counter examples I know to the claim that causes increase the probability of their effects work in this same way. In all cases the cause fails to increase the probability of its effects for the same reason: in the situation described the cause is correlated with some other causal factor which dominates in its effects. This suggests that the condition as stated is too simple. A cause must increase the probability of its effects; but only in situations where such correlations are absent.

The most general situations in which a particular factor is not correlated with any other causal factors are situations in which all other causal factors are held fixed, that is situations that are homogeneous with respect to all other causal factors. In the population where everyone exercises, smoking cannot be correlated with exercising. So, too, in populations where no-one is an exerciser. I hypothesize then that the correct connection between causal laws and laws of association is this:

'C causes E' if and only if C increases the probability of E in every situation which is otherwise causally homogeneous with respect to E.

Carnap's notion of a state description[10] can be used to pick out the causally homogeneous situations. A complete set of causal factors for E is the set of all C_i such that either $C_i \hookrightarrow + E$ or $C_i \hookrightarrow \neg E$. (For short $C_i \hookrightarrow \pm E$.) Every possible arrangement of the factors from a set which is complete except for C picks out a population homogeneous in all causal factors but C. Each such arrangement is given by one of the 2^n state descriptions $K_j = \wedge \pm C_i$ over the set $\{C_i\}$ (i ranging from 1 to n) consisting of all alternative causal factors. These are the only situations in which probabilities tell anything about causal laws. I will refer to them as *test* situations for the law $C \hookrightarrow E$.

Using this notation the connection between laws of association and causal laws is this:

CC: $C \hookrightarrow E$ iff $\text{Prob}(E/C.K_j) > \text{Prob}(E/K_j)$ for all state descriptions K_j over the set $\{C_i\}$, where $\{C_i\}$ satisfies

(i) $C_i \in \{C_i\} \Rightarrow C_i \hookrightarrow \pm E$

(ii) $C \notin \{C_i\}$

(iii) $\forall D (D \hookrightarrow \pm E \Rightarrow D = C$ or $D \in \{C_i\})$

(iv) $C_i \in \{C_i\} \Rightarrow \neg (C \hookrightarrow C_i)$.

Condition (iv) is added to ensure that the state descriptions do not hold fixed any factors in the causal chain from C to E. It will be discussed further in the section after next.

Obviously CC does not provide an analysis of the schema $C \hookrightarrow E$, because exactly the same schema appears on both sides of the equivalence. But it does impose mutual constraints, so that given sets of causal and associational laws cannot be arbitrarily conjoined. CC is, I believe, the strongest connection that can be drawn between causal laws and laws of association.

1.1 Two Advantages for Scientific Explanation

C. G. Hempel's original account of inductive-statistical explanation[11] had two crucial features which have been given up in later accounts, particularly in Salmon's: (1) an

---

[10] See Rudolf Carnap, *The Continuum of Inductive Methods* (Chicago: University of Chicago Press, 1952).

[11] See C. G. Hempel, *Aspects of Scientific Explanation* (New York: Free Press, 1965).

explanatory factor must increase the probability of the fact to be explained; (2) what counts as a good explanation is an objective, person-independent matter. Both of these features seem to me to be right. If we use causal laws in explanations, we can keep both these requirements and still admit as good explanations just those cases that are supposed to argue against them.

(i) Hempel insisted that an explanatory factor increase the probability of the phenomenon it explains. This is an entirely plausible requirement, although there is a kind of explanation for which it is not appropriate. In one sense, to explain a phenomenon is to locate it in a nomic pattern. The aim is to lay out all the laws relevant to the phenomenon; and it is irrelevant to this aim whether the phenomenon has high or low probability under these laws. Although this seems to be the kind of explanation that Richard Jeffrey describes in 'Statistical Explanation vs. Statistical Inference',[12] it is not the kind of explanation that other of Hempel's critics have in mind. Salmon, for instance, is clearly concerned with causal explanation.[13] Even for causal explanation Salmon thinks the explanatory factor may decrease the probability of the factor to be explained. He supports this with the uranium–plutonium example described above.

What makes the uranium count as a good explanation for the clicks in the Geiger counter, however, is not the probabilistic law Salmon cites (Prob(clicks/uranium) < Prob(clicks)), but rather the causal law—'Uranium causes radioactivity'. As required, the probability for radioactive decay increases when the cause is present, *for every test situation*. There is a higher level of radioactivity when uranium is added both for situations in which polonium is present, and for situations in which polonium is absent. Salmon sees the probability

---

[12] See Richard C. Jeffrey, 'Statistical Explanation vs. Statistical Inference', in Wesley Salmon, op. cit.

[13] This is explicitly stated in Salmon's later papers (see 'Theoretical Explanation' in S. Körner, *Explanation* (Oxford: Basil Blackwell, 1975), but it is already clear from the treatment in 'Statistical Explanation' that Salmon is concerned with causal explanations, otherwise there is no accounting for his efforts to rule out 'spurious' correlations as explanatory.

decreasing because he attends to a population which is not causally homogeneous.

Insisting on increase in probability across all test situations not only lets in the good cases of explanation which Salmon cites; it also rules out some bad explanations that must be admitted by Salmon. For example, consider a case which, so far as the law of association is concerned, is structurally similar to Salmon's uranium example. I consider eradicating the poison oak at the bottom of my garden by spraying it with defoliant. The can of defoliant claims that the spray is 90 per cent effective; that is, the probability of a plant's dying given that it is sprayed is .9, and the probability of its surviving is .1. Here in contrast to the uranium case only the probable outcome, and not the improbable, is explained by the spraying. One can explain why some plants died by remarking that they were sprayed with a powerful defoliant; but this will not explain why some survive.[14]

The difference is in the causal laws. In the favourable example, it is true both that uranium causes high levels of radioactivity and that uranium causes low levels of radio-activity. This is borne out in the laws of association. Holding fixed other causal factors for a given level of decay, either high or low, it is more probable that that level will be reached if uranium is added than not. This is not so in the unfavourable case. It is true that spraying with defoliant causes death in plants, but it is not true that spraying also causes survival. Holding fixed other causes of death, spraying with my defoliant will increase the probability of a plant's dying; but holding fixed other causes of survival, spraying with that defoliant will decrease, not increase, the chances of a plant's surviving.

(ii) All these explanations are explanations by appeal to causal laws. Accounts, like Hempel's or Salmon's or Suppes's, which instead explain by appeal to laws of association, are plagued by the reference class problem. All these accounts allow that one factor explains another just in case some privileged statistical relation obtains between them. (For

---

[14] This example can be made exactly parallel to the uranium–polonium case by imagining a situation in which we choose at random between this defoliant and a considerably weaker one that is only 10 per cent effective.

Hempel the probability of the first factor on the second must be high; for Suppes it must be higher than when the second factor is absent; Salmon merely requires that the probabilities be different.) But whether the designated statistical relation obtains or not depends on what reference class one chooses to look in, or on what description one gives to the background situation. Relative to the description that either the uranium or the polonium is drawn at random, the probability of a large number of clicks is lower when the uranium is present than it is otherwise. Relative to the description that polonium and all other radio-active substances are absent, the probability is higher.

Salmon solves this problem by choosing as the privileged description the description assumed in the request for explanation. This makes explanation a subjective matter. Whether the uranium explains the clicks depends on what information the questioner has to hand, or on what descriptions are of interest to him. But the explanation that Hempel aimed to characterize was in no way subjective. What explains what depends on the laws and facts true in our world, and cannot be adjusted by shifting our interest or our focus.

Explanation by causal law satisfies this requirement. Which causal laws are true and which are not is an objective matter. Admittedly certain statistical relations must obtain; the cause must increase the probability of its effect. But no reference class problem arises. In how much detail should we describe the situations in which this relation must obtain? We must include all and only the other causally relevant features. What interests we have, or what information we focus on, is irrelevant.

I will not here offer a model of causal explanation, but certain negative theses follow from my theory. Note particularly that falling under a causal law (plus the existence of suitable initial conditions) is neither necessary nor sufficient for explaining a phenomenon.

It is not sufficient because a single phenomenon may be in the domain of various causal laws, and in many cases it will be a legitimate question to ask, 'Which of these causal factors actually brought about the effect on this occasion?' This problem is not peculiar to explanation by causal law, however. Both Hempel in his inductive–statistical model and Salmon in the statistical relevance account sidestep the issue by requiring that a 'full' explanation cite all the possibly relevant factors, and not select among them.

Conversely, under the plausible assumption that singular causal statements are transitive, falling under a causal law is not necessary for explanation either. This results from the fact that (as CC makes plain) causal laws are not transitive. Hence a phenomenon may be explained by a factor to which it is linked by a sequence of intervening steps, each step falling under a causal law, without there being any causal law that links the explanans itself with the phenomenon to be explained.

1.2   Some Details and Some Difficulties

Before carrying on to Part 2, some details should be noted and some defects admitted.

(a) *Condition (iv)*. Condition (iv) is added to the above characterization to avoid referring to singular causal facts. A test situation for $C \hookrightarrow E$ is meant to pick out a (hypothetical, infinite) population of individuals which are alike in all causal factors for E, except those which on that occasion are caused by C itself. The test situations should not hold fixed factors in the causal chain from C to E. If it did so, the probabilities in the populations where all the necessary intermediate steps occur would be misleadingly high; and where they do not occur, misleadingly low. Condition (iv) is added to except factors caused by C itself from the description of the test situation. Unfortunately it is too strong. For condition (iv) excepts any factor which *may* be caused by C even on those particular occasions when the factor occurs for other reasons. Still, (iv) is the best method I can think of for dealing with this problem, short of introducing singular causal facts, and I let it stand for the nonce.

(b) *Interactions*. One may ask, 'But might it not happen that $\text{Prob}(E/C) > \text{Prob}(E)$ in *all* causally fixed circumstances, and still C not be a cause of E?' I do not know. I am unable to imagine convincing examples in which it occurs; but that is hardly an answer. But one kind of example is clearly taken account of. That is the problem of spurious correlation (sometimes called 'the problem of joint effects'). If two factors E_1 and E_2 are both effects of a third factor C, then it will frequently happen that the probability of the first factor is greater when the second is present than otherwise, over a wide variety of circumstances. Yet we do not want to assert $E_1 \hookrightarrow E_2$. According to principle CC, however, $E_1 \hookrightarrow E_2$ only if $\text{Prob}(E_1/E_2) > \text{Prob}(E_1)$ both when C obtains, and also when C does not obtain. But the story that E_1 and E_2 are joint effects of C provides no warrant for expecting either of these increases.

One may have a worry in the other direction as well. Must a cause increase the probability of its effect in *every* causally fixed situation? Might it not do so in some, but not in all? I think not. Whenever a cause fails to increase the probability of its effect, there must be a reason. Two kinds of reasons seem possible. The first is that the cause may be correlated with other causal factors. This kind of reason is taken account of. The second is that interaction may occur. Two causal factors are interactive if in combination they act like a single causal factor whose effects are different from at least one of the two acting separately. For example, ingesting an acid poison may cause death; so too may the ingestion of an alkali poison. But ingesting both may have no effect at all on survival.

In this case, it seems, there are three causal truths: (1) ingesting acid without ingesting alkali causes death; (2) ingesting alkali without ingesting acid causes death; and (3) ingesting both alkali and acid does not cause death. All three of these general truths should accord with CC.

Treating interactions in this way may seem to trivialize the analysis; anything may count as a cause. Take any factor that behaves sporadically across variation of causal circumstances. May we not count it as a cause by looking at it separately in those situations where the probability increases, and claim it to be in interaction in any case where the probability does not increase? No. There is no guarantee that this can always be done. For interaction is always interaction with some other *causal factor*; and it is not always possible to find some other factor, or conjunction of factors, which obtain just when the probability of E on the factor at issue decreases, and which itself satisfies principle CC relative to all other causal factors.[15] Obviously, considerably more has to be said about interactions; but this fact at least makes it reasonable to hope they can be dealt with adequately, and that the requirement of increase in probability across all causal situations is not too strong.

(c) *0, 1 probabilities and threshold effects.* Principle CC as it stands does not allow $C \hookrightarrow E$ if there is even a single arrangement of other factors for which the probability of E is one, independent of whether C occurs or not. So CC should be amended to read:

$$C \hookrightarrow E \text{ iff } (\forall_j)\{ \text{Prob}(E/C.K_j) > \text{Prob}(E/K_j) \text{ or } \text{Prob}(E/K_j) = 1 = \text{Prob}(E/C.K_j)\} \text{ and } (\exists_j)\{\text{Prob}(E/K_j) \neq 1\}.$$

It is a consequence of the second conjunct that something that occurs universally can be the consequent of no causal laws. The alternative is to let anything count as the cause of a universal fact.

There is also no natural way to deal with threshold effects, if there are any. If the probability of some phenomenon can be raised just so high, and no higher, the treatment as it stands allows no genuine causes for it.

(d) *Time and causation.* CC makes no mention of time. The properties may be time indexed; taking aspirins at t causes relief at $t + \Delta t$, but the ordering of the indices plays no part in the condition. Time ordering is often introduced in statistical analyses of causation to guarantee the requisite asymmetries. Some, for example, take increase in conditional probability as their basis. But the causal arrow is asymmetric, whereas increase in conditional probability is symmetric: $\text{Prob}(E/C) > \text{Prob}(E)$ iff $\text{Prob}(C/E) > \text{Prob}(C)$. This problem does not arise for CC, because the set of alternative causal factors for E will be different from the set of alternative causal factors for C. I take it to be an advantage that my account leaves open the question of backwards causation. I doubt that we shall ever find compelling examples of it; but if there

[15] See section 3.

were a case in which a later factor increased the probability of an earlier one in all test situations, it might well be best to count it a cause.

2. PROBABILITIES IN DECISION THEORY

Standard versions of decision theory require two kinds of information. (1) How desirable are various combinations of goals and strategies and (2) how effective are various strategies for obtaining particular goals. The first is a question of utilities, which I will not discuss. The second is a matter of effectiveness; it is generally rendered as a question about probabilities. We need to know what may roughly be characterized as 'the probability that the goal will result if the strategy is followed.' It is customary to measure effectiveness by the conditional probability. Following this custom, we could define

$!S$ is an *effective strategy* for G iff $\text{Prob}(G/S) > \text{Prob}(G)$.

I have here used the volative mood marker ! introduced by H. P. Grice,[16] to be read 'let it be the case that'. I shall refer to S as *the strategy state*. For example, if we want to know whether the defoliant is effective for killing poison oak, the relevant strategy state is 'a poison oak plant is sprayed with defoliant'. On the above characterization, the defoliant is effective just in case the probability of a plant's dying, given that it has been sprayed, is greater than the probability of its dying given that it has not been sprayed. Under this characterization the distinction between effective and ineffective strategies depends entirely on what laws of association obtain.

But the conditional probability will not serve in this way, a fact that has been urged by Allan Gibbard and William Harper.[17] Harper and Gibbard point out that the increase in conditional probability may be spurious, and

[16] H. P. Grice, 'Some Aspects of Reason', the Immanuel Kant Lectures, Stanford University, 1977.

[17] See Allan Gibbard and William Harper, 'Counterfactuals and Two Kinds of Expected Utility'. Discussion Paper No. 194, Center for Mathematical Studies in Economics and Management Science Northwestern University, January 1976.

that spurious correlations are no grounds for action. Their own examples are somewhat complex because they specifically address a doctrine of Richard Jeffrey's not immediately to the point here. We can illustrate with the TIAA case already introduced. The probability of long life given that one has a TIAA policy is higher than otherwise. But, as the letter says, it would be a poor strategy to buy TIAA in order to increase one's life expectancy.

The problem of spurious correlation in decision theory leads naturally to the introduction of counterfactuals. We are not, the argument goes, interested in how many people have long lives among people insured by TIAA, but rather in the probability that one *would have* a long life if one *were* insured with TIAA. Apt as this suggestion is, it requires us to evaluate the probability of counterfactuals, for which we have only the beginnings of a semantics (via the device of measures over possible worlds)[18] and no methodology, much less an account of why the methodology is suited to the semantics. How do we test claims about probabilities of counterfactuals? We have no answer, much less an answer that fits with our nascent semantics. It would be preferable to have a measure of effectiveness that requires only probabilities over events that can be tested in the actual world in the standard ways. This is what I shall propose.

The Gibbard and Harper example, an example of spurious correlation due to a joint cause, is a special case of a general problem. We saw that the conditional probability will not serve as a mark of causation in situations where the putative cause is correlated with other causal factors. Exactly the same problem arises for effectiveness. For whatever reason the correlation obtains, the conditional probability is not a good measure of effectiveness in any populations where the strategy state is correlated with other factors causally relevant to the goal state. Increase in conditional probability is no mark of effectiveness in situations which are causally heterogeneous. It is necessary, therefore, to make the same

---

[18] See William Harper, Robert Stalnaker, and Glenn Pearce (eds), *University of Western Ontario Series in Philosophy of Science* (Dordrecht: D. Reidel Publishing Co., 1981).

restrictions about test situations in dealing with strategies that we made in dealing with causes:

!S is an *effective strategy* for obtaining G in situation L iff $\text{Prob}(G/S.K_L) > \text{Prob}(G/K_L)$.

Here K_L is the state description true in L, taken over the complete set $\{C_i\}$ of causal factors for G, barring S. But L may not fix a unique state description. For example L may be the situation I am in when I decide whether to smoke or not, and at the time of the decision it is not determined whether I will be an exerciser. In that case we should compare not the actual values $\text{Prob}(G/S.K_L)$ and $\text{Prob}(G/K_L)$, but rather their expected values:

SC: !S is an *effective strategy* for obtaining G in L iff

$$\sum_j \text{Prob}(G/S.K_j)\text{Prob}(K_j) > \sum_j \text{Prob}(G/K_j)\,\text{Prob}(K_j),$$

where j ranges over all K_j consistent with L.[19]

This formula for computing the effectiveness of strategies has several desired features: (1) it is a function of the probability measure, Prob, given by the laws of association in the actual world; and hence calculable by standard methods of statistical inference. (2) It reduces to the conditional probability in cases where it ought. (3) It restores a natural connection between causes and strategies.

(1) *SC* avoids probabilities over counterfactuals. Implications of the arguments presented here for constructing a semantics for probabilities for counterfactuals will be pointed out in section 2.2.

(2) Troubles for the conditional probability arise in cases like the TIAA example in which there is a correlation between the proposed strategy and (other) causal factors for the goal in question. When such correlations are absent, the conditional probability should serve. This follows immediately: when there are no correlations between S and

---

[19] I first derived this formula by reasoning about experiments. I am especially grateful to David Lewis for pointing out that the original formula was mathematically equivalent to the shorter and more intelligible one presented here.

other causal factors, $\mathrm{Prob}(K_j/S) = \mathrm{Prob}(K_j)$; so the left-hand side of SC reduces to $\mathrm{Prob}(G/S)$ in the situation L and the right-hand side to $\mathrm{Prob}(G)$ in L.

(3) There is a natural connection between causes and strategies that should be maintained; if one wants to obtain a goal, it is a good (in the pre-utility sense of good) strategy to introduce a cause for that goal. So long as one holds both the simple view that increase in conditional probability is a sure mark of causation and the view that conditional probabilities are the right measure of effectiveness, the connection is straightforward. The arguments in Part 1 against the simple view of causation break this connection. But SC re-establishes it, for it is easy to see from the combination of CC and SC that if $X \hookrightarrow G$ is true, then $!X$ will be an effective strategy for G in any situation.

2.1.   *Causal Laws and Effective Strategies*

Although SC joins causes and strategies, it is not this connection that argues for the objectivity of *sui generis* causal laws. As we have just seen, one could maintain the connection between causes and strategies, and still hope to eliminate causal laws by using simple conditional probability to treat both ideas. The reason causal laws are needed in characterizing effectiveness is that they pick out the right properties on which to condition. The K_j which are required to characterize effective strategies must range over *all* and *only* the causal factors for G.

It is easy to see, from the examples of Part 1, why the K_j must include *all* the causal factors. If any are left out, cases like the smoking–heart disease example may arise. If exercising is not among the factors which K_j fixes, the conditional probability of heart disease on smoking may be less than otherwise in K_j, and smoking will wrongly appear as an effective strategy for preventing heart disease.

It is equally important that the K_j not include too much. $\{K_j\}$ partitions the space of possible situations. To partition too finely is as bad as not to partition finely enough. Partitioning on an irrelevancy can make a genuine cause look irrelevant, or make an irrelevant factor look like a cause. Earlier discussion of Simpson's paradox shows that this is structurally possible. Any association between two factors C and E can be reversed by finding a third factor which is correlated in the right way with both. When the third factor is a causal factor, the smaller classes are the right ones to use for judging causal relations between C and E. In these, whatever effects the third factor has on E are held fixed in comparing the effects of C versus those of $\neg C$. But when the third factor is causally irrelevant to E—that is, when it *has* no effects on E—there is no reason for it to be held fixed, and holding it fixed gives wrong judgements both about causes and about strategies.

I will illustrate from a real life case.[20] The graduate school at Berkeley was accused of discriminating against women in their admission policies, thus raising the question 'Does being a woman cause one to be rejected at Berkeley?' The accusation appeared to be borne out in the probabilities: the probability of acceptance was much higher for men than for women. Bickel, Hammel, and O'Connell[21] looked at the data more carefully, however, and discovered that this was no longer true if they partitioned by department. In a majority of the eighty-five departments, the probability of admission for women was just about the same as for men, and in some even higher for women than for men. This is a paradigm of Simpson's paradox. Bickel, Hammel and O'Connell accounted for the paradoxical reversal of associations by pointing out that women tended to apply to departments with high rejection rates, so that department by department women were admitted in about the same ratios as men; but across the whole university considerably fewer women, by proportion, are admitted.

This analysis seems to exonerate Berkeley from the charge of discrimination. But only because of the choice of partitioning variable. If, by contrast, the authors had pointed out that the associations reversed themselves when the

---

[20] Roger Rosenkrantz and Persi Diaconis first pointed out to me that the feature of probabilities described here is called 'Simpson's Paradox', and the reference for this example was supplied by Diaconis.

[21] See Peter J. Bickel, Eugene A. Hammel and J. William O'Connell, 'Sex Bias in Graduate Admissions: Data from Berkeley', in William B. Fairley and Frederick Mosteller, *Statistics and Public Policy* (Reading, Mass: Addison-Wesley, 1977).

applicants were partitioned according to their roller skating
ability that would count as no defence.[22] Why is this so?

The difference between the two situations lies in our
antecedent causal knowledge. We know that applying to a
popular department (one with considerably more applicants
than positions) is just the kind of thing that causes rejection.
But without a good deal more detail, we are not prepared
to accept the principle that being a good roller skater causes
a person to be rejected by the Berkeley graduate school,
and we make further causal judgements accordingly. If the
increased probability for rejection among women disappears
when a causal variable is held fixed, the hypothesis of discrimination in admissions is given up; but not if it disappears
only when some causally irrelevant variable is held fixed.

The Berkeley example illustrates the general point: only
partitions by causally relevant variables count in evaluating
causal laws. If changes in probability under causally irrelevant
partitions mattered, almost any true causal law could be
defeated by finding, somewhere, some third variable that
correlates in the right ways to reverse the required association
between cause and effect.

2.2.   Alternative Accounts which Employ 'True
Probabilities' or Counterfactuals

One may object: once all causally relevant factors have been
fixed, there is no harm in finer partitioning by causally
irrelevant factors. Contrary to what is claimed in the remarks
about roller skating and admission rates, further partitioning
will not change the probabilities. There is a difference between true probabilities and observed relative frequencies.
Admittedly it is likely that one can always find some third,
irrelevant, variable which, on the basis of estimates from
finite data, appears to be correlated with both the cause
and effect in just the ways required for Simpson's paradox.
But we are concerned here not with finite frequencies, or
estimates from them, but rather with true probabilities.
You misread the true probabilities from the finite data, and
think that correlations exist where they do not.

[22] William Kruskal discusses the problem of choosing a partition for these data
briefly in correspondence following the Bickel, Hammel, and O'Connell article
referred to in footnote 21.

For this objection to succeed, an explication is required of the idea of a true probability, and this explication must make plausible the claim that partitions by what are pre-analytically regarded as non-causal factors do not result in different probabilities. It is not enough to urge the general point that the best estimate often differs from the true probability; there must in addition be reason to think that that is happening in every case where too-fine partitioning seems to generate implausible causal hypotheses. This is not an easy task, for often the correlations one would want to classify as 'false' are empirically indistinguishable from others that ought to be classified 'true'. The misleading, or 'false', correlations sometimes pass statistical tests of any degree of stringency we are willing to accept as a general requirement for inferring probabilities from finite data. They will often, for example, be stable both across time and across randomly selected samples.

To insist that these stable frequencies are not true probabilities is to give away too much of the empiricist programme. In the original this programme made two assumptions. First, claims about probabilities are grounded only in stable frequencies. There are notorious problems about finite versus infinite ensembles, but at least this much is certain: what probabilities obtain depends in no way, either epistemologically or metaphysically, on what causal assumptions are made. Secondly, causal claims can be reduced completely to probabilistic claims, although further empirical facts may be required to ensure the requisite asymmetries.

I attack only the second of these two assumptions. Prior causal knowledge is needed along with probabilities to infer new causal laws. But I see no reason here to give up the first, and I think it would be a mistake to do so. Probabilities serve many other concerns than causal reasoning and it is best to keep the two as separate as possible. In his *Grammar of Science* Karl Pearson taught that probabilities should be theory free, and I agree. If one wishes nevertheless to mix causation and probability from the start then at least the arguments I have been giving here show some of the constraints that these 'true probabilities' must meet.

Similar remarks apply to counterfactual analyses. One

popular kind of counterfactual analysis would have it that

!S is effective strategy for G in L iff Prob$(S \mathbin{\Box\!\!\rightarrow} G/L) >$
Prob$(\neg S \mathbin{\Box\!\!\rightarrow} G/L)$[23]

The counterfactual and the causal law approach will agree, only if

$$A: \text{Prob}(\alpha \mathbin{\Box\!\!\rightarrow} G/X) = \text{Prob}(G/\alpha.K_x)$$

where K_x is the maximal *causal* description (barring α) consistent with X. Assuming the arguments here are right, condition A provides an adequacy criterion for any satisfactory semantics of counterfactuals and probabilities.

3. HOW SOME WORLDS COULD NOT BE HUME WORLDS[24]

The critic of causal laws will ask, what difference do they make? A succinct way of putting this question is to consider for every world its corresponding Hume world— a world just like the first in its laws of association, its temporal relations, and even in the sequences of events that occur in it. How does the world that has causal laws as well differ from the corresponding Hume world? I have already argued that the two worlds would differ with respect to strategies.

Here I want to urge a more minor point, but one that might go unnoticed: not all worlds could be turned into Hume worlds by stripping away their causal laws. Given the earlier condition relating causal laws and laws of association, many worlds do not have related Hume worlds. In fact no world whose laws of association provide any correlations could be turned into a Hume world. The demonstration is trivial. Assume that a given world has no causal laws for a particular kind of phenomenon E. The earlier condition tells us to test for causes of E by looking for factors that increase the probability of E in maximal causally homogeneous sub-populations. But in the Hume

---

[23] See articles in Harper, op. cit.

[24] I learned the term 'Hume World' from David Lewis. Apparently it originated with Frank Jackson and other Australian philosophers.

world there are no causes, so every sub-population is homogeneous in all causal factors, and the maximal homogeneous population is the whole population. So if there is any C such that $\mathrm{Prob}(E/C) > \mathrm{Prob}(E)$, it will be true that C causes E, and this world will not be a Hume world after all.

Apparently the laws of association underdetermine the causal laws. It is easy to construct examples in which there are two properties, P and Q, which could be used to partition a population. Under the partition into P and $\neg P$, C increases the conditional probability of E in both sub-populations; but under the partition into Q and $\neg Q$, $\mathrm{Prob}(E/C) = \mathrm{Prob}(E)$. So relative to the assumption that P causes E, but Q does not, 'C causes E' is true. It is false relative to the assumption that $Q \hookrightarrow E$, and $P \not\hookrightarrow E$. This suggests that, for a given set of laws of association, any set of causal laws will do. Once some causal laws have been settled, others will automatically follow, but any starting point is as good as any other. This suggestion is mistaken. Sometimes the causal laws are underdetermined by the laws of association, but not always. Some laws of association are compatible with only one set of causal laws. In general laws of association do not entail causal laws: but in particular cases they can. Here is an example.

Consider a world whose laws of association cover three properties, A, B, and C; and assume that the following are implied by the laws of association:

(1) $\mathrm{Prob}(C/A) > \mathrm{Prob}(C)$
(2) $\mathrm{Prob}(C/B \,\&\, A) > \mathrm{Prob}(C/A)$; $\mathrm{Prob}(C/B \,\&\, \neg A) > \mathrm{Prob}(C/\neg A)$
(3) $\mathrm{Prob}(C/B) = \mathrm{Prob}(C)$

In this world, $B \hookrightarrow C$. The probabilities might for instance be those given in Chart 1. From just the probabilistic facts (1), (2), and (3), it is possible to infer that both A and B are causally relevant to C. Assume $B \hookrightarrow \pm C$. Then by (1), $A \hookrightarrow C$, since the entire population is causally homogeneous (barring A) with respect to C and hence counts as a test population for A's effects on C. But if

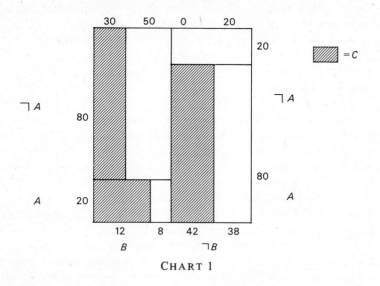

CHART 1

$A \hookrightarrow C$, then by (2), $B \hookrightarrow \pm C$. Therefore $B \hookrightarrow \pm C$. But from (3) this is not possible unless A is also relevant, either positively or negatively, to C. In the particular example pictured in the chart, A and B are both positively relevant to C.

This kind of example may provide solace to the Humean. Often Humeans reject causal laws because they have no independent access to them. They suppose themselves able to determine the laws of association, but they imagine that they never have the initial causal information to begin to apply condition C. If they are lucky, this initial knowledge may not be necessary. Perhaps they live in a world that is not a Hume world; it may nevertheless be a world where causal laws can be inferred just from laws of association.

4. CONCLUSION

The quantity Prob($E/C.K_j$), which appears in both the causal condition of Part 1 and in the measure of effectiveness from Part 2, is called by statisticians the partial conditional probability of E on C, holding K_j fixed; and it is used in ways similar to the ways I have used it here. It forms the foundation

for regression analyses of causation and it is applied by both
Suppes and Salmon to treat the problem of joint effects. In
decision theory the formula *SC* is structurally identical to
one proposed by Brian Skyrms in his deft solution to New-
comb's paradox; and elaborated further in his book *Causal
Necessity*.[25] What is especially significant about the partial
conditional probabilities which appear here is the fact that
these hold fixed all and only causal factors.

The choice of partition, $\{K_j\}$, is the critical feature of the
measure of effectiveness proposed in *SC*. This is both (a)
what makes the formula work in cases where the simple
conditional probability fails; and (b) what makes it necessary to admit causal laws if you wish to sort good strategies
from bad. The way you partition is crucial. In general you
get different results from *SC* if you partition in different
ways. Consider two different partitions for the same space,
$K_1, \ldots, K_n$ and $I_1, \ldots I_s$, which cross-grain each other—
the K_i are mutually disjoint and exhaustive, and so are the
I_j. Then it is easy to produce a measure over the field ($\pm G$,
$\pm C$, $\pm K_i$, $\pm I_j$) such that

$$\sum_{j=1}^{n} \text{Prob}(G/C.K_j)\text{Prob}(K_j) \neq \sum_{j=1}^{n} \text{Prob}(G/C.I_j)\text{Prob}(I_j).$$

What partition is employed is thus essential to whether
a strategy appears effective or not. The right partition—the
one that judges strategies to be effective or ineffective in
accord with what is objectively true—is determined by
what the causal laws are. Partitions by other factors will
give other results; and, if you do not admit causal laws,
there is no general procedure for picking out the right factors.
The objectivity of strategies requires the objectivity of
causal laws.

[25] Brian Skyrms, *Causal Necessity* (New Haven: Yale University Press, 1980).

Essay 2

The Truth Doesn't Explain Much

0. INTRODUCTION

Scientific theories must tell us both what is true in nature, and how we are to explain it. I shall argue that these are entirely different functions and should be kept distinct. Usually the two are conflated. The second is commonly seen as a by-product of the first. Scientific theories are thought to explain by dint of the descriptions they give of reality. Once the job of describing is done, science can shut down. That is all there is to do. To describe nature— to tell its laws, the values of its fundamental constants, its mass distributions—is *ipso facto* to lay down how we are to explain it.

This is a mistake, I shall argue; a mistake that is fostered by the covering-law model of explanation. The covering-law model supposes that all we need to know are the laws of nature—and a little logic, perhaps a little probability theory—and then we know which factors can explain which others. For example, in the simplest deductive–nomological version,[1] the covering-law model says that one factor explains another just in case the occurrence of the second can be deduced from the occurrence of the first given the laws of nature.

But the D-N model is just an example. In the sense which is relevant to my claims here, most models of explanation offered recently in the philosophy of science are covering-law models. This includes not only Hempel's own inductive statistical model,[2] but also Patrick Suppes's probabilistic model of causation,[3] Wesley Salmon's statistical relevance model,[4]

---

[1] See C. G. Hempel, 'Scientific Explanation', in C. G. Hempel (ed.), *Aspects of Scientific Explanation* (New York: Free Press, 1965).

[2] See C. G. Hempel, 'Scientific Explanation', ibid.

[3] See Patrick Suppes, *A Probabilistic Theory of Causality* (Amsterdam: North-Holland Publishing Co., 1970).

[4] See Wesley Salmon, 'Statistical Explanation', in Wesley Salmon (ed.), *Statistical Explanation and Statistical Relevance* (Pittsburgh: University of Pittsburgh Press, 1971).

and even Bengt Hanson's contextualistic model.[5] All these accounts rely on the laws of nature, and just the laws of nature, to pick out which factors we can use in explanation.

A good deal of criticism has been aimed at Hempel's original covering-law models. Much of the criticism objects that these models let in too much. On Hempel's account it seems we can explain Henry's failure to get pregnant by his taking birth control pills, and we can explain the storm by the falling barometer. My objection is quite the opposite. Covering-law models let in too little. With a covering-law model we can explain hardly anything, even the things of which we are most proud—like the role of DNA in the inheritance of genetic characteristics, or the formation of rainbows when sunlight is refracted through raindrops. We cannot explain these phenomena with a covering-law model, I shall argue, because we do not have laws that cover them. Covering laws are scarce.

Many phenomena which have perfectly good scientific explanations are not covered by any laws. No true laws, that is. They are at best covered by *ceteris paribus* generalizations—generalizations that hold only under special conditions, usually ideal conditions. The literal translation is 'other things being equal'; but it would be more apt to read '*ceteris paribus*' as 'other things being *right*.'

Sometimes we act as if this does not matter. We have in the back of our minds an 'understudy' picture of *ceteris paribus* laws: *ceteris paribus* laws are real laws; they can stand in when the laws we would like to see are not available and they can perform all the same functions, only not quite so well. But this will not do. *Ceteris paribus* generalizations, read literally without the '*ceteris paribus*' modifier, are false. They are not only false, but held by us to be false; and there is no ground in the covering-law picture for false laws to explain anything. On the other hand, with the modifier the *ceteris paribus* generalizations may be true, but they cover only those few cases where the conditions are right. For most cases, either we have a law that purports to cover, but cannot explain

[5] See Bengt Hanson, 'Explanations—Of What?' (mimeograph, Stanford University, 1974).

because it is acknowledged to be false, or we have a law that does not cover. Either way, it is bad for the covering-law picture.

1. *CETERIS PARIBUS* LAWS

When I first started talking about the scarcity of covering laws, I tried to summarize my view by saying 'There are no exceptionless generalizations'. Then a friend asked, 'How about "All men are mortal"?' She was right. I had been focusing too much on the equations of physics. A more plausible claim would have been that there are no exceptionless quantitative laws in physics. Indeed not only are there no exceptionless laws, but in fact our best candidates are known to fail. This is something like the Popperian thesis that *every theory is born refuted*. Every theory we have proposed in physics, even at the time when it was most firmly entrenched, was known to be deficient in specific and detailed ways. I think this is also true for every precise quantitative law within a physics theory.

But this is not the point I had wanted to make. Some laws are treated, at least for the time being, as if they were exceptionless, whereas others are not, even though they remain 'on the books'. Snell's law (about the angle of incidence and the angle of refraction for a ray of light) is a good example of this latter kind. In the optics text I use for reference (Miles V. Klein, *Optics*),[6] it first appears on page 21, and without qualification:

Snell's Law: At an interface between dielectric media, there is (also) *a refracted ray* in the second medium, lying in the plane of incidence, making an angle θ_t with the normal, and obeying Snell's law:

$$\sin \theta / \sin \theta_t = n_2/n_1$$

where v_1 and v_2 are the velocities of propagation in the two media, and $n_1 = (c/v_1)$, $n_2 = (c/v_2)$ are the indices of refraction.

---

[6] Miles V. Klein, *Optics* (New York: John Wiley and Sons, 1970), p. 21, italics added. θ is the angle of incidence.

It is only some 500 pages later, when the law is derived from the 'full electromagnetic theory of light', that we learn that Snell's law as stated on page 21 is true only for media whose optical properties are *isotropic*. (In anisotropic media, 'there will generally be *two* transmitted waves'.) So what is deemed true is not really Snell's law as stated on page 21, but rather a refinement of Snell's law:

Refined Snell's Law: *For any two media which are optically isotropic*, at an interface between dielectrics there is a refracted ray in the second medium, lying in the plane of incidence, making an angle θ_t with the normal, such that:

$$\sin\theta/\sin\theta_t = n_2/n_1.$$

The Snell's law of page 21 in Klein's book is an example of a *ceteris paribus* law, a law that holds only in special circumstances—in this case when the media are both isotropic. Klein's statement on page 21 is clearly not to be taken literally. Charitably, we are inclined to put the modifier '*ceteris paribus*' in front to hedge it. But what does this *ceteris paribus* modifier do? With an eye to statistical versions of the covering law model (Hempel's I-S picture, or Salmon's statistical relevance model, or Suppes's probabilistic model of causation) we may suppose that the unrefined Snell's law is not intended to be a universal law, as literally stated, but rather some kind of statistical law. The obvious candidate is a crude statistical law: *for the most part*, at an interface between dielectric media there is *a* refracted ray . . . But this will not do. For *most* media are optically anisotropic, and in an anisotropic medium there are *two* rays. I think there are no more satisfactory alternatives. If *ceteris paribus* laws are to be true laws, there are no statistical laws with which they can generally be identified.

2. WHEN LAWS ARE SCARCE

Why do we keep Snell's law on the books when we both know it to be false and have a more accurate refinement available? There are obvious pedagogic reasons. But are there serious scientific ones? I think there are, and these reasons have to do with the task of explaining. Specifying which factors are explanatorily relevant to which others is a job done by science over and above the job of laying out the laws of nature. Once the laws of nature are known, we still have to decide what kinds of factors can be cited in explanation.

One thing that *ceteris paribus* laws do is to express our explanatory commitments. They tell what kinds of explanations are permitted. We know from the refined Snell's law that in any isotropic medium, the angle of refraction can be explained by the angle of incidence, according to the equation $\sin\theta/\sin\theta_t = n_2/n_1$. To leave the unrefined Snell's law on the books is to signal that the same kind of explanation can be given even for some anisotropic media. The pattern of explanation derived from the ideal situation is employed even where the conditions are less than ideal; and we assume that we can understand what happens in *nearly* isotropic media by rehearsing how light rays behave in pure isotropic cases.

This assumption is a delicate one. It fits far better with the simulacrum account of explanation that I will urge in Essay 8 than it does with any covering-law model. For the moment I intend only to point out that it *is* an assumption, and an assumption which (prior to the 'full electromagnetic theory') goes well beyond our knowledge of the facts of nature. We *know* that in isotropic media, the angle of refraction is due to the angle of incidence under the equation $\sin\theta/\sin\theta_t = n_2/n_1$. We *decide* to explain the angles for the two refracted rays in anisotropic media in the same manner. We may have good reasons for the decision; in this case if the media are nearly isotropic, the two rays will be very close together, and close to the angle predicted by Snell's law; or we believe in continuity of physical processes. But still this decision is not forced by our knowledge of the laws of nature.

Obviously this decision could not be taken if we also had on the books a second refinement of Snell's law, implying that in any anisotropic media the angles are quite different from those given by Snell's law. But laws are scarce, and often we have no law at all about what happens in conditions that are less than ideal.

Covering-law theorists will tell a different story about the use of *ceteris paribus* laws in explanation. From their point of view, *ceteris paribus* explanations are elliptical for genuine covering law explanations from true laws which we do not yet know. When we use a *ceteris paribus* 'law' which we know to be false, the covering-law theorist supposes us to be making a bet about what form the true law takes. For example, to retain Snell's unqualified law would be to bet that the (at the time unknown) law for anisotropic media will entail values 'close enough' to those derived from the original Snell law.

I have two difficulties with this story. The first arises from an extreme metaphysical possibility, in which I in fact believe. Covering-law theorists tend to think that nature is well-regulated; in the extreme, that there is a law to cover every case. I do not. I imagine that natural objects are much like people in societies. Their behaviour is constrained by some specific laws and by a handful of general principles, but it is not determined in detail, even statistically. What happens on most occasions is dictated by no law at all. This is not a metaphysical picture that I urge. My claim is that this picture is as plausible as the alternative. God may have written just a few laws and grown tired. We do not know whether we are in a tidy universe or an untidy one. Whichever universe we are in, the ordinary commonplace activity of giving explanations ought to make sense.

The second difficulty for the ellipsis version of the covering-law account is more pedestrian. Elliptical explanations are not explanations: they are at best assurances that explanations are to be had. The law that is supposed to appear in the complete, correct D-N explanation is not a law we have in our theory, not a law that we can state, let alone test. There may be covering-law explanations in these cases. But those explanations are not our explanations; and those unknown laws cannot be our grounds for saying of a nearly isotropic medium, '$\sin \theta_t \approx k(n_2/n_1)$ *because* $\sin \theta = k$'.

What then are our grounds? I assert only what they are not: they are not the laws of nature. The laws of nature that we know at any time are not enough to tell us what kinds of explanations can be given at that time. That requires a decision; and it is just this decision that covering-law theorists make when they wager about the existence of unknown laws. We may believe in these unknown laws, but we do so on no ordinary grounds: they have not been tested, nor are they derived from a higher level theory. Our grounds for believing in them are only as good as our reasons for adopting the corresponding explanatory strategy, and no better.

3.   WHEN LAWS CONFLICT

I have been maintaining that there are not enough covering laws to go around. Why? The view depends on the picture of science that I mentioned earlier. Science is broken into various distinct domains: hydrodynamics, genetics, laser theory, . . . We have many detailed and sophisticated theories about what happens within the various domains. But we have little theory about what happens in the intersection of domains.

Diagramatically, we have laws like

$$\textit{ceteris paribus, } (x)\,(S(x) \hookrightarrow I(x))$$

and

$$\textit{ceteris paribus, } (x)\,(A(x) \hookrightarrow \neg I(x)).$$

For example, (*ceteris paribus*) adding salt to water decreases the cooking time of potatoes; taking the water to higher altitudes increases it. Refining, if we speak more carefully we might say instead, 'Adding salt to water while keeping the altitude constant decreases the cooking time; whereas increasing the altitude while keeping the saline content fixed increases it'; or

$$(x)\,(S(x) \,\&\, \neg A(x) \hookrightarrow I(x))$$

and

$$(x)\,(A(x) \,\&\, \neg S(x) \hookrightarrow \neg I(x)).$$

But neither of these tells what happens when we both add salt to the water and move to higher altitudes.

Here we think that probably there is a precise answer

about what would happen, even though it is not part of our common folk wisdom. But this is not always the case. I discuss this in detail in the next essay. Most real life cases involve some combination of causes; and general laws that describe what happens in these complex cases are not always available. Although both quantum theory and relativity are highly developed, detailed, and sophisticated, there is no satisfactory theory of relativistic quantum mechanics. A more detailed example from transport theory is given in the next essay. The general lesson is this: where theories intersect, laws are usually hard to come by.

4. WHEN EXPLANATIONS CAN BE GIVEN ANYWAY

So far, I have only argued half the case. I have argued that covering laws are scarce, and that *ceteris paribus* laws are no true laws. It remains to argue that, nevertheless, *ceteris paribus* laws have a fundamental explanatory role. But this is easy, for most of our explanations are explanations from *ceteris paribus* laws.

Let me illustrate with a humdrum example. Last year I planted camellias in my garden. I know that camellias like rich soil, so I planted them in composted manure. On the other hand, the manure was still warm, and I also know that camellia roots cannot take high temperatures. So I did not know what to expect. But when many of my camellias died, despite otherwise perfect care, I knew what went wrong. The camellias died because they were planted in hot soil.

This is surely the right explanation to give. Of course, I cannot be absolutely certain that this explanation is the correct one. Some other factor may have been responsible, nitrogen deficiency or some genetic defect in the plants, a factor that I did not notice, or may not even have known to be relevant. But this uncertainty is not peculiar to cases of explanation. It is just the uncertainty that besets all of our judgements about matters of fact. We must allow for oversight; still, since I made a reasonable effort to eliminate other menaces to my camellias, we may have some confidence that this is the right explanation.

So we have an explanation for the death of my camellias. But it is not an explanation from any true covering law. There is no law that says that camellias just like mine, planted in soil which is both hot and rich, die. To the contrary, they do not all die. Some thrive; and probably those that do, do so *because* of the richness of the soil they are planted in. We may insist that there must be some differentiating factor which brings the case under a covering law: in soil which is rich and hot, camellias of one kind die; those of another thrive. I will not deny that there may be such a covering law. I merely repeat that our ability to give this humdrum explanation precedes our knowledge of that law. On the Day of Judgment, when all laws are known, these may suffice to explain all phenomena. But in the meantime we do give explanations; and it is the job of science to tell us what kinds of explanations are admissible.

In fact I want to urge a stronger thesis. If, as is possible, the world is not a tidy deterministic system, this job of telling how we are to explain will be a job which is still left when the descriptive task of science is complete. Imagine for example (what I suppose actually to be the case) that the facts about camellias are irreducibly statistical. Then it is possible to know all the general nomological facts about camellias which there are to know—for example, that 62 per cent of all camellias in just the circumstances of my camellias die, and 38 per cent survive.[7] But one would not thereby know how to explain what happened in my garden. You would still have to look to the *Sunset Garden Book* to learn that the *heat* of the soil explains the perishing, and the *richness* explains the plants that thrive.

5. CONCLUSION

Most scientific explanations use *ceteris paribus* laws. These laws, read literally as descriptive statements, are false, not only false but deemed false even in the context of use. This

[7] Various writers, especially Suppes (footnote 3) and Salmon (footnote 4), have urged that knowledge of more sophisticated statistical facts will suffice to determine what factors can be used in explanation. I do not believe that this claim can be carried out, as I have argued in Essay 1.

is no surprise: we want laws that unify; but what happens may well be varied and diverse. We are lucky that we can organize phenomena at all. There is no reason to think that the principles that best organize will be true, nor that the principles that are true will organize much.

Essay 3

Do the Laws of Physics State the Facts?

0. INTRODUCTION

There is a view about laws of nature that is so deeply entrenched that it does not even have a name of its own. It is the view that laws of nature describe facts about reality. If we think that the facts described by a law obtain, or at least that the facts that obtain are sufficiently like those described in the law, we count the law true, or true-for-the-nonce, until further facts are discovered. I propose to call this doctrine the *facticity* view of laws. (The name is due to John Perry.)

It is customary to take the fundamental explanatory laws of physics as the ideal. Maxwell's equations, or Schroedinger's, or the equations of general relativity, are paradigms, paradigms upon which all other laws—laws of chemistry, biology, thermodynamics, or particle physics—are to be modelled. But this assumption confutes the facticity view of laws. For the fundamental laws of physics do not describe true facts about reality. Rendered as descriptions of facts, they are false; amended to be true, they lose their fundamental, explanatory force.

To understand this claim, it will help to contrast biology with physics. J. J. C. Smart argues that biology has no genuine laws of its own.[1] It resembles engineering. Any general claim about a complex system, such as a radio or a living organism, will be likely to have exceptions. The generalizations of biology, or engineering's rules of thumb, are not true laws because they are not exceptionless. Many (though not Smart himself) take this to mean that biology is a second-rate science. If this is good reasoning, it must be physics that is the second-rate science. Not only do the laws of physics have exceptions; unlike biological laws, they are not even true for the most part, or approximately true.

[1] See J. J. C. Smart, *Philosophy and Scientific Realism* (London: Routledge and Keegan Paul, 1963).

The view of laws with which I begin——'Laws of nature describe facts about reality'——is a pedestrian view that, I imagine, any scientific realist will hold. It supposes that laws of nature tell how objects of various kinds behave: how they behave some of the time, or all of the time, or even (if we want to prefix a necessity operator) how they must behave. What is critical is that they talk about objects—— real concrete things that exist here in our material world, things like quarks, or mice, or genes; and they tell us what these objects do.

Biological laws provide good examples. For instance, here is a generalization taken from a Stanford text on chordates:

The gymnotoids [American knife fish] are slender fish with enormously long anal fins, which suggest the blade of a knife of which the head is a handle. They often swim slowly with the body straight by un- dulating this fin. They [presumably 'always' or 'for the most part'] are found in Central and South America . . . Unlike the characins they ['usually'?] hide by day under river banks or among roots, or even bury themselves in sand, emerging only at night.[2]

The fundamental laws of physics, by contrast, do not tell what the objects in their domain do. If we try to think of them in this way, they are simply false, not only false but deemed false by the very theory that maintains them. But if physics' basic, explanatory laws do not describe how things behave, what do they do? Once we have given up facticity, I do not know what to say. Richard Feynman, in *The Character of Physical Law*, offers an idea, a metaphor. Feynman tells us 'There is . . . a rhythm and a pattern be- tween the phenomena of nature which is not apparent to the eye, but only to the eye of analysis; and it is these rhythms and patterns which we call Physical Laws . . .'[3] Most philosophers will want to know a lot more about how these rhythms and patterns function. But at least Feynman does not claim that the laws he studies describe the facts.

I say that the laws of physics do not provide true descriptions

[2] R. McNeill Alexander, *The Chordates* (Cambridge: Cambridge University Press, 1975), p. 179.
[3] Richard Feynman, *The Character of Physical Law* (Cambridge, Mass: MIT Press, 1967), p. 13.

of reality. This sounds like an anti-realist doctrine. Indeed it is, but to describe the claim in this way may be misleading. For anti-realist views in the philosophy of science are traditionally of two kinds. Bas van Fraassen[4] is a modern advocate of one of these versions of anti-realism; Hilary Putnam[5] of the other. Van Fraassen is a sophisticated instrumentalist. He worries about the existence of unobservable entities, or rather, about the soundness of our grounds for believing in them; and he worries about the evidence which is supposed to support our theoretical claims about how these entities behave. But I have no quarrel with theoretical entities; and for the moment I am not concerned with how we know what they do. What is troubling me here is that our explanatory laws do not tell us what they do. It is in fact part of their explanatory role not to tell.

Hilary Putnam in his version of internal realism also maintains that the laws of physics do not represent facts about reality. But this is because nothing—not even the most commonplace claim about the cookies which are burning in the oven—represents facts about· reality. If anything did, Putnam would probably think that the basic equations of modern physics did best. This is the claim that I reject. I think we can allow that all sorts of statements represent facts of nature, including the generalizations one learns in biology or engineering. It is just the fundamental explanatory laws that do not truly represent. Putnam is worried about meaning and reference and how we are trapped in the circle of words. I am worried about truth and explanation, and how one excludes the other.

1.   EXPLANATION BY COMPOSITION OF CAUSES, AND THE TRADE-OFF OF TRUTH AND EXPLANATORY POWER

Let me begin with a law of physics everyone knows—the law of universal gravitation. This is the law that Feynman

---

[4] See Bas van Fraassen, *The Scientific Image* (Oxford: Clarendon Press, 1980).

[5] See Hilary Putnam, *Meaning and the Moral Sciences* (London: Routledge and Kegan Paul, 1978) and 'Models and Reality', *Journal of Symbolic Logic*, forthcoming.

uses for illustration; he endorses the view that this law is 'the greatest generalization achieved by the human mind'.[6]

$$\text{Law of Gravitation: } F = Gmm'/r^2.$$

In words, Feynman tells us:

The Law of Gravitation is that two bodies exert a force between each other which varies inversely as the square of the distance between them, and varies directly as the product of their masses.[7]

Does this law truly describe how bodies behave?

Assuredly not. Feynman himself gives one reason why. 'Electricity also exerts forces inversely as the square of the distance, this time between charges . . .'[8] It is not true that for *any* two bodies the force between them is given by the law of gravitation. Some bodies are charged bodies, and the force between them is not Gmm'/r^2. Rather it is some resultant of this force with the electric force to which Feynman refers.

For bodies which are both massive and charged, the law of universal gravitation and Coulomb's law (the law that gives the force between two charges) interact to determine the final force. But neither law by itself truly describes how the bodies behave. No charged objects will behave just as the law of universal gravitation says; and any massive objects will constitute a counterexample to Coulomb's law. These two laws are not true; worse, they are not even approximately true. In the interaction between the electrons and the protons of an atom, for example, the Coulomb effect swamps the gravitational one, and the force that actually occurs is very different from that described by the law of gravity.

There is an obvious rejoinder: I have not given a complete statement of these two laws, only a shorthand version. The Feynman version has an implicit *ceteris paribus* modifier in front, which I have suppressed. Speaking more carefully, the law of universal gravitational is something like this:

[6] Feynman, op. cit., p. 14.
[7] Ibid., p. 14.
[8] Ibid., p. 30.

If there are no forces other than gravitational forces at work, *then* two bodies exert a force between each other which varies inversely as the square of the distance between them, and varies directly as the product of their masses.

I will allow that this law is a true law, or at least one that is held true within a given theory. But it is not a very useful law. One of the chief jobs of the law of gravity is to help explain the forces that objects experience in various complex circumstances. *This* law can explain in only very simple, or ideal, circumstances. It can account for why the force is as it is when just gravity is at work; but it is of no help for cases in which both gravity and electricity matter. Once the *ceteris paribus* modifier has been attached, the law of gravity is irrelevant to the more complex and interesting situations.

This unhappy feature is characteristic of explanatory laws. I said that the fundamental laws of physics do not represent the facts, whereas biological laws and principles of engineering do. This statement is both too strong and too weak. Some laws of physics do represent facts, and some laws of biology—particularly the explanatory laws—do not. The failure of facticity does not have so much to do with the nature of physics, but rather with the nature of explanation. We think that nature is governed by a small number of simple, fundamental laws. The world is full of complex and varied phenomena, but these are not fundamental. They arise from the interplay of more simple processes obeying the basic laws of nature. (Later essays will argue that even simple isolated processes do not in general behave in the uniform manner dictated by fundamental laws.)

This picture of how nature operates to produce the subtle and complicated effects we see around us is reflected in the explanations that we give: we explain complex phenomena by reducing them to their more simple components. This is not the only kind of explanation we give, but it is an important and central kind. I shall use the language of John Stuart Mill, and call this *explanation by composition of causes.*[9]

It is characteristic of explanations by composition of

[9] John Stuart Mill, *A System of Logic* (New York: Harper and Brothers, 1893). See Book III, Chapter VI.

causes that the laws they employ fail to satisfy the requirement of facticity. The force of these explanations comes
from the presumption that the explanatory laws 'act' in
combination just as they would 'act' separately. It is critical,
then, that the laws cited have the same form, in or out of
combination. But this is impossible if the laws are to describe
the actual behaviour of objects. The actual behaviour is the
resultant of simple laws in combination. The effect that
occurs is not an effect dictated by any one of the laws
separately. In order to be true in the composite case, the
law must describe one effect (the effect that actually happens);
but to be explanatory, it must describe another. There is
a trade-off here between truth and explanatory power.

2. HOW VECTOR ADDITION INTRODUCES CAUSAL POWERS

Our example, where gravity and electricity mix, is an example
of the composition of forces. We know that forces add
vectorially. Does vector addition not provide a simple and
obvious answer to my worries? When gravity and electricity
are both at work, two forces are produced, one in accord
with Coulomb's law, the other according to the law of
universal gravitation. Each law is accurate. Both the gravitational and the electric force are produced as described;
the two forces then add together vectorially to yield the
total 'resultant' force.

The vector addition story is, I admit, a nice one. But it
is just a metaphor. *We* add forces (or the numbers that
represent forces) when we do calculations. Nature does not
'add' forces. For the 'component' forces are not there,
in any but a metaphorical sense, to be added; and the laws
that say they are there must also be given a metaphorical
reading. Let me explain in more detail.

The vector addition story supposes that Feynman has
left something out in his version of the law of gravitation.
In the way that he writes it, it sounds as if the law describes
the *resultant* force exerted between two bodies, rather
than a component force—the force which is *produced
between the two bodies in virtue of their gravitational*

masses (or, for short, the force *due to gravity*). A better way to state the law would be

Two bodies produce a force between each other (the force due to gravity) which varies inversely as the square of the distance between them, and varies directly as the product of their masses.

Similarly, for Coulomb's law

Two charged bodies produce a force between each other (the force due to electricity) which also varies inversely as the square of the distance between them, and varies directly as the product of their masses.

These laws, I claim, do not satisfy the facticity requirement. They appear, on the face of it, to describe what bodies do: in the one case, the two bodies produce a force of size Gmm'/r^2; in the other, they produce a force of size qq'/r^2. But this cannot literally be so. For the force of size Gmm'/r^2 and the force of size qq'/r^2. are not real, occurrent forces. In interaction a single force occurs—the force we call the 'resultant'—and this force is neither the force due to gravity nor the electric force. On the vector addition story, the gravitational and the electric force are both produced, yet neither exists.

Mill would deny this. He thinks that in cases of the composition of causes, each separate effect does exist—it exists as *part* of the resultant effect, just as the left half of a table exists as part of the whole table. Mill's paradigm for composition of causes is mechanics. He says:

In this important class of cases of causation, one cause never, properly speaking, defeats or frustrates another; both have their full effect. If a body is propelled in two directions by two forces, one tending to drive it to the north, and the other to the east, it is caused to move in a given time exactly as far in *both* directions as the two forces would separately have carried it . . .[10]

Mill's claim is unlikely. Events may have temporal parts, but not parts of the kind Mill describes. When a body has moved along a path due north-east, it has travelled neither due north nor due east. The first half of the motion can be a part of the total motion; but no pure north motion can be

[10] Ibid., Bk. III, Ch. VI.

a part of a motion that always heads northeast. (We learn this from Judith Jarvis Thomson's *Acts and Other Events*.) The lesson is even clearer if the example is changed a little: a body is pulled equally in opposite directions. It does not budge, but in Mill's picture it has been caused to move both several feet to the left and several feet to the right. I realize, however, that intutions are strongly divided on these cases; so in the next section I will present an example for which there is no possibility of seeing the separate effects of the composed causes as part of the effect which actually occurs.

It is implausible to take the force due to gravity and the force due to electricity literally as parts of the actually occurring force. Is there no way to make sense of the story about vector addition? I think there is, but it involves giving up the facticity view of laws. We can preserve the truth of Coulomb's law and the law of gravitation by making them about something other than the facts: the laws can describe the causal powers that bodies have.

Hume taught that 'the distinction, which we often make betwixt *power* and the *exercise* of it, is . . . without foundation'.[11] It is just Hume's illicit distinction that we need here: the law of gravitation claims that two bodies have the *power* to produce a force of size Gmm'/r^2. But they do not always succeed in the *exercise* of it. What they actually produce depends on what other powers are at work, and on what compromise is finally achieved among them. This may be the way we do sometimes imagine the composition of causes. But if so, the laws we use talk not about what bodies do, but about the powers they possess.

The introduction of causal powers will not be seen as a very productive starting point in our current era of moderate empiricism. Without doubt, we do sometimes think in terms of causal powers, so it would be foolish to maintain that the facticity view must be correct and the use of causal powers a total mistake. But facticity cannot be given up easily. We need an account of what laws are, an account that connects them, on the one hand, with standard scientific methods

[11] David Hume, *A Treatise of Human Nature*, ed. L. A. Selby Bigge (Oxford: Clarendon Press, 1978), p. 311.

for confirming laws, and on the other, with the use they are put to for prediction, construction, and explanation. If laws of nature are presumed to describe the facts, then there are familiar, detailed philosophic stories to be told about why a sample of facts is relevant to their confirmation, and how they help provide knowledge and understanding of what happens in nature. Any alternative account of what laws of nature do and what they say must serve at least as well; and no story I know about causal powers makes a very good start.

3.   THE FORCE DUE TO GRAVITY

It is worth considering further *the force due to gravity* and *the force due to electricity*, since this solution is frequently urged by defenders of facticity. It is one of a class of suggestions that tries to keep the separate causal laws in something much like their original form, and simultaneously to defend their facticity by postulating some intermediate effect which they produce, such as a force due to gravity, a gravitational potential, or a field.

Lewis Creary has given the most detailed proposal of this sort that I know. Creary claims that there are two quite different kinds of laws that are employed in explanations where causes compose—laws of causal influence and laws of causal action. Laws of causal influence, such as Coulomb's law and the law of gravity, 'tell us what forces or other causal influences operate in various circumstances', whereas laws of causal action 'tell us what the results are of such causal influences, acting either singly or in various combinations'.[12] In the case of composition of forces, the law of interaction is a vector addition law, and vector addition laws 'permit explanations of an especially satisfying sort' because the analysis 'not only identifies the different component causal influences at work, but also quantifies their relative importance'.[13] Creary also describes less satisfying kinds of composition, including reinforcement, interference,

---

[12] Lewis Creary, 'Causal Explanation and the Reality of Natural Component Forces,' *Pacific Philosophical Quarterly* 62(1981), p. 153.

[13] Ibid., p. 153.

and predomination. On Creary's account, Coulomb's law and the law of gravity come out true because they correctly describe what influences are produced—here, the force due to gravity and the force due to electricity. The vector addition law then combines the separate influences to predict what motions will occur.

This seems to me to be a plausible account of how a lot of causal explanation is structured. But as a defence of the truth of fundamental laws, it has two important drawbacks. First, in many cases there are no *general* laws of interaction. Dynamics, with its vector addition law, is quite special in this respect. This is not to say that there are no truths about how this specific kind of cause combines with that, but rather that theories can seldom specify a procedure that works from one case to another. Without that, the collection of fundamental laws loses the generality of application which Creary's proposal hoped to secure. The classical study of irreversible processes provides a good example of a highly successful theory that has this failing. Flow processes like diffusion, heat transfer, or electric current ought to be studied by the transport equations of statistical mechanics. But usually, the model for the distribution functions and the details of the transport equations are too complex: the method is unusable. A colleague of mine in engineering estimates that 90 per cent of all engineering systems cannot be treated by the currently available methods of statistical mechanics. 'We analyze them by whatever means seem appropriate for the problem at hand,' he adds.[14]

In practice engineers handle irreversible processes with old fashioned phenomenological laws describing the flow (or flux) of the quantity under study. Most of thse laws have been known for quite a long time. For example there is Fick's law, dating from 1855, which relates the diffusion velocity of a component in a mixture to the gradient of its density ($J_m = - D\partial c/\partial x$). Equally simple laws describe other processes: Fourier's law for heat flow, Newton's law for sheering force (momentum flux) and Ohm's law for electric current. Each of these is a linear differential equation

---

[14] Kline's studies of methods of approximation appear in S. J. Kline, *Similitude and Approximation Theory* (New York: McGraw-Hill, 1969), p. 140.

in t (e.g. the J_m in Fick's law cited above is dm/dt), giving the time rate of change of the desired quantity (in the case of Fick's law, the mass). Hence a solution at one time completely determines the quantity at any other time. Given that the quantity can be controlled at some point in a process, these equations should be perfect for determining the future evolution of the process. They are not.

The trouble is that each equation is a *ceteris paribus* law. It describes the flux only so long as just one kind of cause is operating. More realistic examples set different forces at play simultaneously. In a mixture of liquids, for example, if both the temperatures and the concentrations are non-uniform, there may be a flow of liquid due not only to the concentration gradients but also to the temperature gradients. This is called the Soret effect.

The situation is this. For the several fluxes J we have laws of the form,

$$J_m = f_1(\alpha_m)$$
$$\cdot$$
$$\cdot$$
$$\cdot$$
$$J_q = f_n(\alpha_q)$$

Each of these is appropriate only when its α is the only relevant variable. For cross-effects we require laws of the form.

$$J_m = g_1(\alpha_m, \ldots, \alpha_q)$$
$$\cdot$$
$$\cdot$$
$$J_q = g_n(\alpha_m, \ldots, \alpha_q).$$

This case is structurally just like the simple causal examples that I discussed in the last essay. We would like to have laws that combine different processes. But we have such laws only in a few special cases, like the Soret effect. For the Soret effect we assume simple linear additivity in our law of action, and obtain a final cross-effect law by adding a thermal diffusion factor into Fick's law. But this law of causal action is highly specific to the situation and will not work for combining arbitrary influences studied by transport theory.

Are there any principles to be followed in modifying to allow for cross-effects? There is only one systematic account of cross-effect modification for flow processes. It originated with Onsager in 1931, but was not developed until the 1950s. Onsager theory defines force-flux pairs, and prescribes a method for writing cross-effect equations involving different forces. As C. A. Truesdell describes it, 'Onsagerism claims to unify and correlate much existing knowledge of irreversible processes'.[15] Unfortunately it does not succeed. Truesdell continues:

As far as concerns heat conduction, viscosity, and diffusion . . . this is not so. Not only does Onsagerism not apply to any of these phenomena without a Procrustean force-fit, but even in the generous interpretation of its sectaries it does not yield as much reduction for the theory of viscosity as was known a century earlier and follows from fundamental principles. . . .[16]

Truesdell claims that the principles used in Onsager theory are vacuous. The principles must sometimes be applied in one way, sometimes in another, in an *ad hoc* fashion demanded by each new situation. The prescription for constructing laws, for example, depends on the proper choice of conjugate flux-force pairs. Onsager theory offers a general principle for making this choice, but if the principle were followed literally, we would not make the proper choice in even the most simple situations. In practice on any given occasion the choice is left to the physicist's imagination. It seems that after its first glimmer of generality the Onsager approach turns out to be a collection of *ad hoc* techniques.

I have illustrated with statistical mechanics; but this is not a special case. In fact classical mechanics may well be the only discipline where a general law of action is always available. This limits the usefulness of Creary's idea. Creary's scheme, if it works, buys facticity, but it is of little benefit to realists who believe that the phenomena of nature flow from a small number of abstract, fundamental laws. The fundamental laws will be severely limited in scope. Where

---

[15] C. A. Truesdell, *Rational Thermodynamics* (New York: McGraw-Hill, 1969), p. 140.
[16] Ibid., p. 140.

the laws of action go case by case and do not fit a general scheme, basic laws of influence, like Coulomb's law and the law of gravity, may give true accounts of the influences that are produced; but the work of describing what the influences do, and what behaviour results, will be done by the variety of complex and ill-organized laws of action: Fick's law with correction factors, and the like. This fits better with my picture of a nature best described by a vast array of phenomenological laws tailored to specific situations, than with one governed in an orderly way from first principles.

The causal influences themselves are the second big drawback to Creary's approach. Consider our original example. Creary changes it somewhat from the way I originally set it up. I had presumed that the aim was to explain the size and direction of a resultant force. Creary supposes that it is not a resultant force but a consequent motion which is to be explained. This allows him to deny the reality of the resultant force. We are both agreed that there cannot be three forces present—two components and a resultant. But I had assumed the resultant, whereas Creary urges the existence of the components.

The shift in the example is necessary for Creary. His scheme works by interposing an intermediate factor—the causal influence—between the cause and what initially looked to be the effect. In the dynamic example the restructuring is plausible. Creary may well be right about resultant and component forces. But I do not think this will work as a general strategy, for it proliferates influences in every case. Take any arbitrary example of the composition of causes: two laws, where each accurately dictates what will happen when it operates in isolation, say 'C causes E' and C' causes E''; but where C and C' in combination produce some different effect, E''. If we do not want to assume that all three effects—E, E', E''—occur (as we would if we thought that E and E' were parts of E''), then on Creary's proposal we must postulate some further occurrences, F and F', as the *proper* effects of our two laws, effects that get combined by a law of action to yield E'' at the end. In some concrete cases the strategy will work, but in general I see no reason to think that these intermediate influences can always be found. I am not opposed to them because of any general objection to theoretical entities, but rather because I think every new theoretical entity which is admitted should be grounded in experimentation, which shows up its causal structure in detail. Creary's influences seem to me just to be shadow occurrences which stand in for the effects we would like to see but cannot in fact find.

4. A REAL EXAMPLE OF THE COMPOSITION OF CAUSES

The ground state of the carbon atom has five distinct energy levels (see Figure 3.1). Physics texts commonly treat this phenomenon sequentially, in three stages. I shall follow the discussion of Albert Messiah in Volume II of *Quantum Mechanics*.[17] In the first stage, the ground state energy is calculated by a central field approximation; and the single line (*a*) is derived. For some purposes, it is accurate to assume that only this level occurs. But some problems

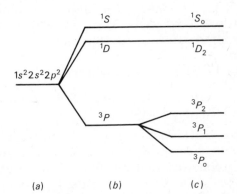

FIG. 3.1. The levels of the ground state of the carbon atom; (*a*) in the central field approximation ($V_1 = V_2 = 0$); (*b*) neglecting spin–orbit coupling ($V_2 = 0$); (*c*) including spin–orbit coupling. (*Source:* Messiah, *Quantum Mechanics*.)

[17] Albert Messiah, *Quantum Mechanics* (Amsterdam: North-Holland, 1961), p. 703.

require a more accurate description. This can be provided by noticing that the central field approximation takes account only of the *average* value of the electrostatic repulsion of the inner shell electrons on the two outer electrons. This defect is remedied at the second stage by considering the effects of a term which is equal to the difference between the exact Coulomb interaction and the average potential used in stage one. This corrective potential 'splits' the single line (*a*) into three lines depicted in (*b*).

But the treatment is inaccurate because it neglects spin effects. Each electron has a spin, or internal angular momentum, and the spin of the electron couples with its orbital angular momentum to create an additional potential. The additional potential arises because the spinning electron has an intrinsic magnetic moment, and 'an electron moving in [an electrostatic] potential "sees" a magnetic field'.[18] About the results of this potential Messiah tells us, 'Only the ^3P state is affected by the spin-orbit energy term; it gets split into three levels: $^3P_0$, $^3P_1$ and $^3P_2$'.[19] Hence the five levels pictured in (*c*).

The philosophic perplexities stand out most at the last stage. The five levels are due to a combination of a Coulomb potential, and a potential created by spin-orbit coupling 'splits' the lowest of these again into three. *That* is the explanation of the five levels. But how can we state the laws that it uses?

For the Coulomb effect we might try

Whenever a Coulomb potential is like that in the carbon atom, the three energy levels pictured in (*b*) occur.

(The real law will of course replace 'like that in the carbon atom' by a mathematical description of the Coulomb potential in carbon; and similarly for 'the three energy levels pictured in (*b*)'.) The carbon atom itself provides a counter-example to this law. It has a Coulomb potential of the right kind; yet the five levels of (*c*) occur, not the three levels of (*b*).

---

[18] Ibid., p. 552.
[19] Ibid., p. 552.

We might, in analogy with the vector addition treatment of composite forces, try instead

The energy levels produced by a Coulomb potential like that in the carbon atom are the three levels pictured in (*b*).

But (as with the forces 'produced by gravity' in our earlier example) the levels that are supposed to be produced by the Coulomb potential are levels that do not occur. In actuality five levels occur, and they do not include the three levels of (*b*). In particular, as we can see from Messiah's diagram, the lowest of the three levels—the ^3P—is not identical with any of the five. In the case of the composition of motions, Mill tried to see the 'component' effects as parts of the actual effect. But that certainly will not work here. The ^3P level in (*b*) may be 'split' and hence 'give rise to', the $^3P_0$, $^3P_1$, and $^3P_2$ levels in (*c*); but it is certainly not a part of any of these levels.

It is hard to state a true factual claim about the effects of the Coulomb potential in the carbon atom. But quantum theory does guarantee that a certain *counterfactual* is true; the Coulomb potential, if it *were* the only potential at work, would produce the three levels in (*b*). Clearly this counterfactual bears on our explanation. But we have no model of explanation that shows how. The covering-law model shows how statements of fact are relevant to explaining a phenomenon. But how is a truth about energy levels, which would occur in quite different circumstances, relevant to the levels which do occur in these? We think the counterfactual is important; but we have no account of how it works.

5. COMPOSITION OF CAUSES VERSUS EXPLANATION BY COVERING LAW

The composition of causes is not the only method of explanation which can be employed. There are other methods, and some of these are compatible with the facticity view of laws. Standard covering-law explanations are a prime example.

Sometimes these other kinds of explanation are available

even when we give an explanation which tells what the component causes of a phenomenon are. For example, in the case of Coulomb's law and the law of gravity, we know how to write down a more complex law (a law with a more complex antecedent) which says exactly what happens when a system has both mass and charge. Mill thinks that such 'super' laws are always available for mechanical phenomena. In fact he thinks, 'This explains why mechanics is a deductive or demonstrative science, and chemistry is not'.[20]

I want to make three remarks about these super laws and the covering explanations they provide. The first is familiar from the last essay: super laws are not always available. Secondly, even when they are available, they often do not explain much. Thirdly, and most importantly, even when other good explanations are to hand, if we fail to describe the component processes that go together to produce a phenomenon, we lose a central and important part of our understanding of what makes things happen.

(1) There are a good number of complex scientific phenomena which we are quite proud to be able to explain. As I urged in the last essay, for many of these explanations, super covering laws are not available to us. Sometimes we have every reason to believe that a super law exists. In other cases we have no good empirical reason to suppose even this much. Nevertheless, after we have seen what occurs in a specific case, we are often able to understand how various causes contributed to bring it about. We do explain, even without knowing the super laws. We need a philosophical account of explanations which covers this very common scientific practice, and which shows why these explanations are good ones.

(2) Sometimes super laws, even when they are available to cover a case, may not be very explanatory. This is an old complaint against the covering-law model of explanation: 'Why does the quail in the garden bob its head up and down in that funny way whenever it walks?' . . . 'Because they all do.' In the example of spin-orbit coupling

[20] Mill, op. cit., p. 267.

it does not explain the five energy levels that appear in a particular experiment to say 'All carbon atoms have five energy levels'.

(3) Often, of course, a covering law for the complex case will be explanatory. This is especially true when the antecedent of the law does not just piece together the particular circumstances that obtain on the occasion in question, but instead gives a more abstract description which fits with a general body of theory. In the case of spin-orbit coupling, Stephen Norman remarks that quantum mechanics provides general theorems about symmetry groups, and Hamiltonians, and degeneracies, from which we could expect to derive, covering-law style, the energy levels of carbon from the appropriate abstract characterization of its Hamiltonian, and the symmetries it exhibits.

Indeed we can do this; and if we do not do it, we will fail to see that the pattern of levels in carbon is a particular case of a general phenomenon which reflects a deep fact about the effects of symmetries in nature. On the other hand, to do only this misses the detailed causal story of *how* the splitting of spectral lines by the removal of symmetry manages to get worked out in each particular case.

This two-faced character is a widespread feature of explanation. Even if there is a single set of super laws which unifies all the complex phenomena one studies in physics, our current picture may yet provide the ground for these laws: what the unified laws dictate should happen, happens *because* of the combined action of laws from separate domains, like the law of gravity and Coulomb's law. Without these laws, we would miss an essential portion of the explanatory story. Explanation by subsumption under super, unified covering laws would be no replacement for the composition of causes. It would be a complement. To understand how the consequences of the unified laws are brought about would require separate operation of the law of gravity, Coulomb's law, and so forth; and the failure of facticity for these contributory laws would still have to be faced.

6.   CONCLUSION

There is a simple, straightforward view of laws of nature which is suggested by scientific realism, the facticity view: laws of nature describe how physical systems behave. This is by far the commonest view, and a sensible one; but it does not work. It does not fit explanatory laws, like the fundamental laws of physics. Some other view is needed if we are to account for the use of laws in explanation; and I do not see any obvious candidate that is consistent with the realist's reasonable demand that laws describe reality and state facts that might well be true. There is, I have argued, a trade-off between factual content and explanatory power. We explain certain complex phenomena as the result of the interplay of simple, causal laws. But what do these laws *say*? To play the role in explanation we demand of them, these laws must have the same form when they act together as when they act singly. In the simplest case, the consequences that the laws prescribe must be exactly the same in interaction, as the consequences that would obtain if the law were operating alone. But then, what the law states cannot literally be true, for the consequences that would occur if it acted alone are not the consequences that actually occur when it acts in combination.

If we state the fundamental laws as laws about what happens when only a single cause is at work, then we can suppose the law to provide a true description. The problem arises when we try to take that law and use it to explain the very different things which happen when several causes are at work. This is the point of 'The Truth Doesn't Explain Much'. There is no difficulty in writing down laws which we suppose to be true: '*If* there are no charges, no nuclear forces, . . . *then* the force between two masses of size m and m' separated by a distance r is Gmm'/r^2.' We count this law true—what it says will happen, does happen—or at least happens to within a good approximation. But this law does not explain much. It is irrelevant to cases where there are electric or nuclear forces at work. The laws of physics, I concluded, to the extent that they are true, do not explain much. We could know all the true laws of nature, and still not know how to explain composite cases. Explanation must rely on something other than law.

But this view is absurd. There are not two vehicles for explanation: laws for the rare occasions when causes occur separately; and another secret, nameless device for when they occur in combination. Explanations work in the same way whether one cause is at work, or many. 'Truth Doesn't Explain' raises perplexities about explanation by composition of causes; and it concludes that explanation is a very peculiar scientific activity, which commonly does not make use of laws of nature. But scientific explanations do use laws. It is the laws themselves that are peculiar. The lesson to be learned is that the laws that explain by composition of causes fail to satisfy the facticity requirement. If the laws of physics are to explain how phenomena are brought about, they cannot state the facts.

Essay 4

The Reality of Causes
in a World of Instrumental Laws

0. INTRODUCTION

Empiricists are notoriously suspicious of causes. They have
not been equally wary of laws. Hume set the tradition when
he replaced causal facts with facts about generalizations.
Modern empiricists do the same. But nowadays Hume's
generalizations are the laws and equations of high level
scientific theories. On current accounts, there may be some
question about where the laws of our fundamental theories
get their necessity; but it is no question that these laws
are the core of modern science. Bertrand Russell is well
known for this view:

> The law of gravitation will illustrate what occurs in any exact science . . .
> Certain differential equations can be found, which hold at every in-
> stant for every particle of the system . . . But there is nothing that
> could be properly called 'cause' and nothing that could be properly
> called 'effect' in such a system.[1]

For Russell, causes 'though useful to daily life and in the
infancy of a science, tend to be displaced by quite different
laws as soon as a science is successful'.
 · It is convenient that Russell talks about physics, and that
the laws he praises are its fundamental equations—Hamilton's
equations or Schroedinger's, or the equations of general
relativity. That is what I want to discuss too. But I hold
just the reverse of Russell's view. I am in favour of causes
and opposed to laws. I think that, given the way modern
theories of mathematical physics work, it makes sense only
to believe their causal claims and not their explanatory laws.

---

[1] Bertrand Russell, 'On the Notion of Cause', *Mysticism and Logic* (London:
Allen & Unwin, 1917), p. 194.

1. EXPLAINING BY CAUSES

Following Bromberger, Scriven, and others, we know that
there are various things one can be doing in explaining.
Two are of importance here: in explaining a phenomenon
one can cite the causes of that phenomenon; or one can set
the phenomenon in a general theoretical framework. The
framework of modern physics is mathematical, and good
explanations will generally allow us to make quite precise
calculations about the phenomena we explain. Rene Thom
remarks the difference between these two kinds of explanation, though he thinks that only the causes really explain:
'DesCartes with his vortices, his hooked atoms, and the
like explained everything and calculated nothing; Newton,
with the inverse square of gravitation, calculated everything
and explained nothing'.[2]

Unlike Thom, I am happy to call both explanation, so long
as we do not illicitly attribute to theoretical explanation
features that apply only to causal explanation. There is a
tradition, since the time of Aristotle, of deliberately conflating the two. But I shall argue that they function quite
differently in modern physics. If we accept Descartes's
causal story as adequate, we must count his claims about
hooked atoms and vortices true. But we do not use Newton's
inverse square law as if it were either true or false.

One powerful argument speaks against my claim and for
the truth of explanatory laws—the *argument from coincidence*. Those who take laws seriously tend to subscribe
to what Gilbert Harman has called inference to the best
explanation. They assume that the fact that a law *explains*
provides evidence that the law is true. The more diverse
the phenomena that it explains, the more likely it is to be
true. It would be an absurd coincidence if a wide variety
of different kinds of phenomena were all explained by a
particular law, and yet were not in reality consequent from
the law. Thus the argument from coincidence supports a
good many of the inferences we make to best explanations.

The method of inference to the best explanation is subject

[2] Rene Thom, *Structural Stability and Morphogenesis*, trans. C. H. Wadding-
ton (Reading, Mass.: W. A. Benjamin, 1972), p. 5.

to an important constraint, however—the requirement of non-redundancy. We can infer the truth of an explanation only if there are no alternatives that account in an equally satisfactory way for the phenomena. In physics nowadays, I shall argue, an acceptable causal story is supposed to satisfy this requirement. But exactly the opposite is the case with the specific equations and models that make up our theoretical explanations. There is redundancy of theoretical treatment, but not of causal account.

There is, I think, a simple reason for this: causes make their effects happen. We begin with a phenomenon which, relative to our other general beliefs, we think would not occur unless something peculiar brought it about. In physics we often mark this belief by labelling the phenomena as effects—the Sorbet effect, the Zeeman effect, the Hall effect. An effect needs something to bring it about, and the peculiar features of the effect depend on the particular nature of the cause, so that—in so far as we think we have got it right—we are entitled to infer the character of the cause from the character of the effect.

But equations do not bring about the phenomenological laws we derive from them (even if the phenomenological laws are themselves equations). Nor are they used in physics as if they did. The specific equations we use to treat particular phenomena provide a way of casting the phenomena into the general framework of the theory. Thus we are able to treat a variety of disparate phenomena in a similar way, and to make use of the theory to make quite precise calculations. For both of these purposes it is an advantage to multiply theoretical treatments.

Pierre Duhem used the redundancy requirement as an argument against scientific realism, and recently Hilary Putman uses it as an argument against realism in general. Both propose that, in principle, for any explanation of any amount of data there will always be an equally satisfactory alternative. The history of science suggests that this claim may be right: we constantly construct better explanations to replace those of the past. But such arguments are irrelevant here; they do not distinguish between causal claims and theoretical accounts. Both are likely to be replaced by better accounts in the future.

Here I am not concerned with alternatives that are at best available only in principle, but rather with the practical availability of alternatives within theories we actually have to hand. For this discussion, I want to take the point of view that Putnam calls 'internal realism'; to consider actual physical theories which we are willing to account as acceptable, even if only for the time being, and to ask, 'Relative to that theory, which of its explanatory claims are we to deem true?' My answer is that causal claims are to be deemed true, but to count the basic explanatory laws as true is to fail to take seriously how physics succeeds in giving explanations.

I will use two examples to show this. The first—quantum damping and its associated line broadening—is a phenomenon whose understanding is critical to the theory of lasers. Here we have a single causal story, but a fruitful multiplication of successful theoretical accounts. This contrasts with the unacceptable multiplication of causal stories in the second example.

There is one question we should consider before looking at the examples, a question pressed by two colleagues in philosophy of science, Dan Hausman and Robert Ennis. How are we to distinguish the explanatory laws, which I argue are not to be taken literally, from the causal claims and more pedestrian statements of fact, which are? The short answer is that there is no way. A typical way of treating a problem like this is to find some independent criterion —ideally syntactical, but more realistically semantical— which will divide the claims of a theory into two parts. Then it is argued that claims of one kind are to be taken literally, whereas those of the other kind function in some different way.

This is not what I have in mind. I think of a physics theory as providing an explanatory scheme into which phenomena of interest can be fitted. I agree with Duhem here. The scheme simplifies and organizes the phenomena so that we can treat similarly happenings that are phenomenologically different, and differently ones that are phenomenologically the same. It is part of the nature of this organizing activity that it cannot be done very well if we stick too closely to stating what is true. Some claims of the theory must be literally descriptive (I think the claims about the mass and charge of the electron are a good example) if the theory is to be brought to bear on the phenomena; but I suspect that there is no general independent way of characterizing which these will be. What is important to realize is that if the theory is to have considerable explanatory power, most of its fundamental claims will not state truths, and that this will in general include the bulk of our most highly prized laws and equations.

2. EXAMPLES: QUANTUM DAMPING

In radiative damping, atoms de-excite, giving off photons whose frequencies depend on the energy levels of the atom. We know by experiment that the emission line observed in a spectroscope for a radiating atom is not infinitely sharp, but rather has a finite linewidth; that is, there is a spread of frequencies in the light emitted. What causes this natural linewidth? Here is the standard answer which physicists give, quoted from a good textbook on quantum radiation theory by William Louisell:

There are many interactions which may broaden an atomic line, but the most fundamental one is the reaction of the radiation field on the atom. That is, when an atom decays spontaneously from an excited state radiatively, it emits a quantum of energy into the radiation field. This radiation may be reabsorbed by the atom. The reaction of the field on the atom gives the atom a linewidth and causes the original level to be shifted as we show. This is the source of the natural linewidth and the Lamb shift.[3]

Following his mathematical treatment of the radiative decay, Louisell continues:

We see that the atom is continually emitting and reabsorbing quanta of radiation. The energy level shift does not require energy to be conserved while the damping requires energy conservation. Thus damping

[3] William H. Louisell, *Quantum-Statistical Properties of Radiation* (New York: John Wiley & Sons, 1973), p. 285.

is brought about by the emission and absorption of real photons while the photons emitted and absorbed which contribute to the energy shift are called virtual photons.[4]

This account is universally agreed upon. Damping, and its associated line broadening, are brought about by the emission and absorption of real photons.

Here we have a causal story; but not a mathematical treatment. We have not yet set line broadening into the general mathematical framework of quantum mechanics. There are many ways to do this. One of the Springer Tracts by G. S. Agarwal[5] summarizes the basic treatments which are offered. He lists six different approaches in his table of contents: (1) Weisskopf–Wigner method; (2) Heitler–Ma method; (3) Goldberger–Watson method; (4) Quantum Statistical method: master equations; (5) Langevin equations corresponding to the master equation and a c-number representation; and (6) neoclassical theory of spontaneous emission.

Before going on to discuss these six approaches, I will give one other example. The theory of damping forms the core of current quantum treatments of lasers. Figure 4.1 is a diagram from a summary article by H. Haken on 'the' quantum theory of the laser.[6] We see that the situation I described for damping theory is even worse here. There are so many different treatments that Haken provides a 'family tree' to set straight their connections. Looking at the situation Haken himself describes it as a case of 'theory overkill'. Laser theory is an extreme case, but I think there is no doubt that this kind of redundancy of treatment, which Haken and Agarwal picture, is common throughout physics.

Agarwal describes six treatments of line broadening. All six provide precise and accurate calculations for the shape

[4] Ibid., p. 289.

[5] See G. S. Agarwal, *Quantum-Statistical Theories of Spontaneous Emission and their Relation to Other Approaches* (Berlin: Springer-Verlag, 1974).

[6] H. Haken, 'The Semiclassical and Quantum Theory of the Laser', in S. M. Kay and A. Maitland (eds), *Quantum Optics* (London: Academic Press, 1970), p. 244.

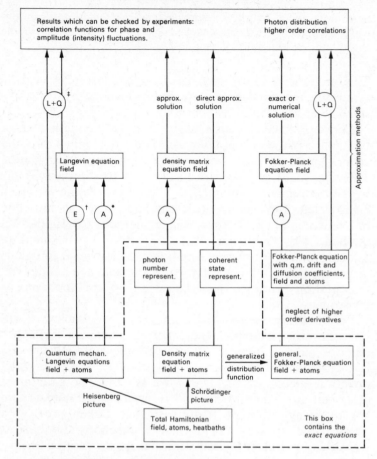

*A : adiabatic elimination of atomic variables
† E : exact elimination of atomic variables
‡ L + Q: linearization and quantum mechanical quasilinearization

FIG. 4.1. Family tree of the quantum theory of the laser. (*Source*:
   Haken, 'The Semiclassical and Quantum Theory of the Laser'.)

and width of the broadened line. How do they differ? All of
the approaches employ the basic format of quantum mechanics. Each writes down a Schroedinger equation; but it is
a *different* equation in each *different* treatment. (Actually
among the six treatments there are really just three different

equations.) The view that I am attacking takes theoretical explanations to provide, as best they can, statements of objective laws. On this view the six approaches that Agarwal lists compete with one another; they offer different laws for exactly the same phenomena.

But this is not Agarwal's attitude. Different approaches are useful for different purposes; they complement rather than compete. The Langevin and Master equations of (4) and (5), for instance, have forms borrowed from statistical mechanics. They were introuduced in part because the development of lasers created an interest in photon correlation experiments. Clearly, if we have statistical questions, it is a good idea to start with the kind of equations from which we know how to get statistical answers.

Let us consider an objection to the point of view I have been urging. We all know that physicists write down the kinds of equations they know how to solve; if they cannot use one approximation, they try another; and when they find a technique that works, they apply it in any place they can. These are commonplace observations that remind us of the pragmatic attitude of physicists. Perhaps, contrary to my argument, the multiplication of theoretical treatments says more about this pragmatic orientation than it does about how explanatory laws ought to be viewed. I disagree. I think that it does speak about laws, and in particular shows how laws differ from causes. We do not have the same pragmatic tolerance of causal alternatives. We do not use first one causal story in explanation, then another, depending on the ease of calculation, or whatever.

The case of the radiometer illustrates. The radiometer was introduced by William Crookes in 1873, but it is still not clear what makes it work. Recall from the Introduction to these essays that there are three plausible theories. The first attributes the motion of the vanes to light pressure. This explanation is now universally rejected. As M. Goldman remarks in 'The Radiometer Revisited',

A simple calculation shows that on a typical British summer day, when the sky is a uniform grey (equally luminous all over) the torque from the black and silver faces exactly balance, so that for a perfect radiometer [i.e., a radiometer with a perfect vacuum] no motion would be possible.[7]

Two explanations still contend. The first is the more standard, textbook account, which is supported by Goldman's calculations. It supposes that the motion is caused by the perpendicular pressure of the gas in the perfect vacuum against the vanes. But as we have seen, on Maxwell's account the motion must be due to the tangential stress created by the gas slipping around the edge of the vanes. There is a sense in which Maxwell and Goldman may both be right: the motion may be caused by a combination of tangential and perpendicular stress. But this is not what they claim. Each claims that the factor he cites is the single significant factor in bringing about the motion, and only one or the other of these claims can be accepted. This situation clearly constrasts with Agarwal's different theoretical treatments. In so far as we are interested in giving a causal explanation of the motion, we must settle on one account or the other. We cannot use first one account, then the other, according to our convenience.

I know of this example through Francis Everitt, who thinks of building an experiment that would resolve the question. I mention Everitt's experiment again because it argues for the difference in objectivity which I urge between theoretical laws and causal claims. It reminds us that unlike theoretical accounts, which can be justified only by an inference to the best explanation, causal accounts have an independent test of their truth: we can perform controlled experiments to find out if our causal stories are right or wrong. Experiments of these kinds in fact play an important role in an example from which Wesley Salmon defends inferences to the best explanation.

3.   THE ARGUMENT FROM COINCIDENCE

In a recent paper[8] Salmon considers Jean Perrin's arguments for the existence of atoms and for the truth of Avagadro's

---

[7] M. Goldman, 'The Radiometer Revisited', *Physics Education* 13 (1978), p. 428.   [8] See footnote 21, Introduction.

hypothesis that there are a fixed number of molecules in any gram mole of a fluid. Perrin performed meticulous experiments on Brownian motion in colloids from which he was able to calculate Avagadro's number quite precisely. His 1913 tract, in which he summarizes his experiments and recounts the evidence for the existence of atoms, helped sway the community of physicists in favour of these hypotheses. Besides Brownian motion, Perrin lists thirteen quite different physical situations which yield a determination of Avogadro's number. So much evidence of such a variety of kinds all pointing to the same value must surely convince us, urges Perrin, that atoms exist and that Avogadro's hypothesis is true.

For many, Perrin's reasoning is a paradigm of inference to the best explanation; and it shows the soundness of that method. I think this misdiagnoses the structure of the argument. Perrin does not make an inference to the best explanation, where explanation includes anything from theoretical laws to a detailed description of how the explanandum was brought about. He makes rather a more restricted inference —an inference to the most probable cause.

A well-designed experiment is constructed to allow us to infer the character of the cause from the character of its more readily observable effects. Prior to Perrin, chemists focused their attention on the size and velocities of the suspended particles. But this study was unrewarding; the measurements were difficult and the results did not signify much. Perrin instead studied the height distribution of the Brownian granules at equilibrium. From his results, with a relatively simple model for the collision interactions, he was able to calculate Avogadro's number. Perrin was a brilliant experimenter. It was part of his genius that he was able to find quite specific effects which were peculiarly sensitive to the exact character of the causes he wanted to study. Given his model, the fact that the carrier fluids had just 6×10^{23} atoms for every mole made precise and calculable differences to the distribution he observed.

The role of the model is important. It brings out exactly what part coincidence plays in the structure of Perrin's argument. Our reasoning from the character of the effect to the character of the cause is always against a background of other knowledge. We aim to find out about a cause with a particular structure. What effects appear as a result of that structure will be highly sensitive to the exact nature of the causal processes which connect the two. If we are mistaken about the processes that link cause and effect in our experiment, what we observe may not result in the way we think from the cause under study. Our results may be a mere artefact of the experiment, and our conclusions will be worthless.

Perrin explicitly has this worry about the first óf the thirteen phenomena he cites: the viscosity of gases, which yields a value for Avogadro's number via Van der Waal's equation and the kinetic theory of gases. In his *Atoms* he writes that 'the probable error, for all these numbers is roughly 30 per cent, owing to the approximations made in the calculations that lead to the Clausius-Maxwell and Van der Waal's equations.' He continues: 'The Kinetic Theory justly excites our admiration. [But] it fails to carry complete conviction, because of the many hypotheses it involves.' (I take it he means 'unsubstantiated hypotheses'.) What sets Perrin's worries to rest? He tell us himself in the next sentence: 'If by entirely independent routes we are led to the same values for the molecular magnitudes, we shall certainly find our faith in the theory considerably strengthened.'[9]

Here is where coincidence enters. We have thirteen phenomena from which we can calculate Avogadro's number. Any one of these phenomena—if we were sure enough about the details of how the atomic behaviour gives rise to it—would be good enough to convince us that Avogadro is right. Frequently we are not sure enough; we want further assurance that we are observing genuine results and not experimental artefacts. This is the case with Perrin. He lacks confidence in some of the models on which his calculations are based. But he can appeal to coincidence. Would it not be a coincidence if each of the observations was an artefact, and yet all agreed so closely about Avogadro's number?

[9] Jean Perrin, *Atoms*, trans. D. Ll. Hammick (New York: D. Van Nostrand Co., 1916), p. 82.

The convergence of results provides reason for thinking that the various models used in Perrin's diverse calculations were each good enough. It thus reassures us that those models can legitimately be used to infer the nature of the cause from the character of the effects.

In each of Perrin's thirteen cases we infer a concrete cause from a concrete effect. We are entitled to do so because we assume that causes make effects occur in just the way that they do, via specific, concrete causal processes. The structure of the cause physically determines the structure of the effect. Coincidence enters Perrin's argument, but not in a way that supports inference to the best explanation in general. There is no connection analogous to causal propagation between theoretical laws and the phenomenological generalizations which they bring together and explain. Explanatory laws summarize phenomenological laws; they do not make them true. Coincidence will not help with laws. We have no ground for inferring from any phenomenological law that an explanatory law must be just so; multiplying cases cannot help.

I mentioned that Gilbert Harman introduced the expression 'inference to the best explanation'. Harman uses two examples in his original paper.[10] The first is the example that we have just been discussing: coming to believe in atoms. The second is a common and important kind of example from everyday life: inferring that the butler did it. Notice that these are both cases in which we infer facts about concrete causes: they are not inferences to the laws of some general explanatory scheme. Like Perrin's argument, these do not vindicate a general method for inferring the truth of explanatory laws. What they illustrate is a far more restrictive kind of inference: inference to the best cause.

4. CONCLUSION

Perrin did not make an inference to the best explanation, only an inference to the most probable cause. This is typical of modern physics. 'Competing' theoretical

[10] G. H. Harman, 'Inference to the Best Explanation', *Philosophical Review* 74 (1965), pp. 88–95.

treatments—treatments that write down different laws for the same phenomena—are encouraged in physics, but only a single causal story is allowed. Although philosophers generally believe in laws and deny causes, explanatory practice in physics is just the reverse.

Essay 5

When Explanation Leads to Inference

0. INTRODUCTION

When can we infer the best explanation? This question divides scientific realists on the one hand, from operationalists, instrumentalists, positivists, and constructive empiricists on the other. There must obviously be certain provisions to ensure that 'the best' is good enough. But once these are understood, the realist answer to the question is 'always'; the anti-realist, 'never'. The realist asks, 'How *could* something explain if it was not true?' The anti-realist thinks this question exposes a mistaken view about what we do in explaining. Explanations (at least the high level explanations of theoretical science which are the practical focus of the debate) organize, briefly and efficiently, the unwieldy, and perhaps unlearnable, mass of highly detailed knowledge that we have of the phenomena. But organizing power has nothing to do with truth.

I am going to discuss two anti-realists, Bas van Fraassen and Pierre Duhem. Van Fraassen's book, *The Scientific Image*,[1] provides a powerful and elegant defence of a brand of anti-realism which he calls 'constructive empiricism'. Duhem's views are laid out in his classic work of 1906, *The Aim and Structure of Physical Theory*.[2] According to van Fraassen the constructive empiricist maintains:

Science aims to give us theories which are empirically adequate; and acceptance of a theory involves as belief only that it is empirically adequate. (Basically, a theory is empirically adequate 'exactly if what it says about the observable things and events in this world is true'.)[3]

Van Fraassen presents the difference between the realist and the constructive empiricist as one of attitude. Both may

---

[1] Bas C. van Fraassen, *The Scientific Image* (Oxford: Clarendon Press, 1980).
[2] See Pierre Duhem, *The Aim and Structure of Physical Theory*, trans. Philip P. Wiener (New York: Atheneum, 1962).
[3] van Fraassen, op. cit., p. 12.

explain by showing how the phenomena at hand can be derived from certain fundamental princples. But the two kinds of philosopher have opposing attitudes to the principles. The realist believes that they are true and genuinely give rise to the phenomena; the constructive empiricist believes merely that the principles are sufficient to derive the phenomena.

The realist, says van Fraassen, is making a mistake. When a theory succeeds in saving the phenomena, the scientific realist is ready to infer that its laws are true (or near true, or true for the nonce) and that its entities exist. Van Fraassen holds that a theory's success at saving the phenomena gives reason to believe *just that*: that it saves the phenomena, and nothing more. That the theory is true is a gratuitous additional assumption.

This is the core of Duhem's view as well. Duhem has no quarrel with phenomenological laws, which can be confirmed by inductive methods. What he opposes are theoretical laws, whose only ground is their ability to explain. Like van Fraassen, Duhem rejects theoretical laws because he does not countenance inference to the best explanation. Neither van Fraassen nor Duhem are opposed to ampliative inference in general. They make a specific and concrete attack on a particular kind of inference which they see as invalid—inference to the best explanation—and thereby on the scientific realism to which it gives rise.

This is the real interest of their view. They have a specific objection to one mode of reasoning and one class of scientific conclusions. They do not argue from a sweeping sceptical position that starts with the weakness of our senses and the poverty of our capacities and concludes that no one can ever know anything. Nor do they argue from a theory of meaning that counts theoretical talk as devoid of truth value along with all of our claims to morality, causality, and religion. Finally, they are not transcendental idealists like Kant. Nor (to use Ian Hacking's apt label) are they transcendental nominalists like Hilary Putnam, who argues that since thought can never connect with reality, our knowledge can achieve at best an internal coherence. Duhem and van Fraassen make a distinction within the field of scientific knowledge, while scepticism, positivism, and the transcendentalisms are global doctrines about the whole domain of science. Duhem and van Fraassen allow that many inferences are sound, but not inferences to pure theory that are justified only in terms of explanation.

Their arguments are persuasive. But I think that van Fraassen and Duhem eliminate more than they should. It is apparent from earlier essays that I share their anti-realism about theoretical laws. On the other hand, I believe in theoretical entities, and that is my main topic in this essay. Arguments against inference to the best explanation do not work against the explanations that theoretical entities provide. These are causal explanations, and inference from effect to cause is legitimate. I will have nothing new to say about the structure of these inferences. I aim only to show that we can be realists about theoretical entities on van Fraassen and Duhem's very own grounds.

1.   VAN FRAASSEN'S ATTACK

Van Fraassen asks, 'Why should I believe in theoretical entities?' There is a canonical answer: there are no genuine regularities at the phenomenological level. It is only among theoretical entities that science finds true regularities. Once we have laid these out, we have a powerful explanatory scheme. The exceptionless laws that we postulate at the theoretical level can explain not only why the phenomena are as regular as they are, but why we see the exceptions we do. Van Fraassen grants this. But, he asks, what reason do we have for inferring from the fact that a bundle of principles save the phenomena to the fact that they are true? We need some reason, some good reason, though certainly not a conclusive reason. Many arguments wear their validity on their sleeve: 'I think. Therefore, I exist.' But not, 'P explains Q. Q is true. Therefore P is true.'

This argument, and Duhem's as well, assumes that truth is an external characteristic of explanation; i.e., that something could satisfy all the other criteria for being an explanation and yet fail to be true. This is the way we are often taught to think of Ptolemaic astronomy. It might well form a completely satisfactory explanatory scheme, yet that does not settle the question of its truth. This, for instance, is what the medieval Piccolomini, one of Duhem's heroes, says of Ptolemy and his successors:

for these astronomers it was amply sufficient that their constructs save the appearances, that they allow for the reckoning of the movements of the heavenly bodies, their arrangements, and their place. Whether or not things really are as they envisage them—that question they leave to the philosophers of nature.[4]

As we saw in the last essay, Duhem's own argument against inference to the best explanation is the argument from redundancy: for any given set of phenomena, in principle there will always be more than one equally satisfactory explanation, and some of these explanations will be incompatible. Since not all of them can be true, it is clear that truth is independent of satisfactoriness for explanation. Sometimes Duhem's argument is read epistemologically instead. He is taken to make a point, not about what the criteria are, but rather about our ability to see if they obtain. On the epistemological reading Duhem maintains merely that there will always be different laws which appear equally true, *so far as we can ever tell*, and yet are incompatible.

This, I think, is a mistaken reading. For it is a general feature of our knowledge and does not show what is peculiar to the inference-to-best-explanation which Duhem attacks. Duhem is not, for example, opposed to phenomenological laws, which arise by inductive generalization. It is a familiar fact that it is possible to construct different inductive rules which give rise to different generalizations from the same evidence. Here too there will always be more than one incompatible law which appears equally true so far as we can tell. These kinds of problems with inductive inference were known to Duhem. But he did not dwell on them. His concern was not with epistemological issues of this sort, but rather with the relationship between truth and explanation.

I said that Duhem and van Fraassen take truth to be an

---

[4] Pierre Duhem, *To Save the Phenomena*, trans. Edmund Doland and Chanenah Maschler (Chicago: University of Chicago Press, 1969), p. 82.

external characteristic to explanation. Here is an analogy. I ask you to tell me an interesting story, and you do so. I may add that the story should be true. But if I do so, that is a new, additional requirement like the one which Piccolomini's natural philosophers make. They will call something a *genuine* explanation only if it does all else it should and is *in addition* true. Here is another analogy. Van Fraassen and Duhem challenge us to tell what is special about the explanatory relation. Why does the truth of the second relatum guarantee the truth of the first? Two-placed relations do not in general have that characteristic. Consider another two-placed relation: —— is the paper at the 1980 Western Washington Conference that immediately preceded ——. In that year my paper immediately preceded Richard Wollheim's; Wollheim's paper may well have been true. But that does not make mine true.

2.   THE CASE FOR THEORETICAL ENTITIES

Van Fraassen and Duhem argue that explanation has truth going along with it only as an extra ingredient. But causal explanations have truth built into them. When I infer from an effect to a cause, I am asking what made the effect occur, what brought it about. No explanation of that sort explains at all unless it does present a cause; and in accepting such an explanation, I am accepting not only that it explains in the sense of organizing and making plain, but also that it presents me with a cause. My newly planted lemon tree is sick, the leaves yellowing and dropping off. I finally explain this by saying that water has accumulated in the base of the planter: the water is the cause of the disease. I drill a hole in the base of the oak barrel where the lemon tree lives, and foul water flows out. That was the cause. Before I had drilled the hole, I could still give the explanation and to give that explanation was to present the supposed cause, the water. There must *be* such water for the explanation to be correct. An explanation of an effect by a cause has an existential component, not just an optional extra ingredient.

Likewise when I explain the change in rate in fall of a

a light droplet in an electric field, by asserting that there are positrons or electrons on the ball, I am inferring from effect to cause, and the explanation has no sense at all without the direct implication that there are electrons or positrons on the ball. Here there is no drilling a hole to let the electrons gush out before our eyes. But there is the generation of other effects: if the ball is negatively charged, I spray it with a positron emitter and thereby change the rate of fall of the ball: positrons from the emitter wipe out the electrons on the ball. What I invoke in completing such an explanation are not fundamental laws of nature, but rather properties of electrons and positrons, and highly complex, highly specific claims about just what behaviour they lead to in just this situation. I infer to the best explanation, but only in a derivative way: I infer to the most probable cause, and that cause is a specific item, what we call a theoretical entity. But note that the electron is not an entity of any particular theory. In a related context van Fraassen asks if it is the Bohr electron, the Rutherford electron, the Lorenz electron or what. The answer is, it is the electron, about which we have a large number of incomplete and sometimes conflicting theories.

Indeed I should use an example of van Fraassen's here to show how we differ. In a cloud chamber we see certain tracks which he says have roughly the same physical explanation as the vapour trail from a jet in the sky. In each case I may explain the trail by stating some laws. But what about the entities? I say that the most probable cause of the track in the chamber is a particle, and as I find out more, I can even tell you with some specificity what kind of particle. That, argues van Fraassen, is still quite different from the jet in the sky. For, there, van Fraassen says, look at the speck just ahead of the trail, or here, use this powerful glass to spy it out. There is no such spying out when we get to the cloud chamber. I agree to that premise but not to the conclusion. In explaining the track by the particle, I am saying that the particle causes the track, and that explanation, or inference to the most probable cause, has no sense unless one is asserting that the particle in motion brings about, causes, makes, produces, that very track. The particle in the cloud chamber is just one example. Our belief in theoretical entities is generally founded on inferences from concrete effects to concrete causes. Here there is an answer to the van Fraassen–Duhem question. What is special about explanation by theoretical entity is that it is causal explanation, and existence is an internal characteristic of causal claims. There is nothing similar for theoretical laws.

Van Fraassen does not believe in causes. He takes the whole causal rubric to be a fiction. That is irrelevant here. Someone who does not believe in causes will not give causal explanations. One may have doubts about some particular causal claims, or, like van Fraassen, about the whole enterprise of giving causal explanations. These doubts bear only on how satisfactory you should count a causal explanation. They do not bear on what kind of inferences you can make once you have accepted that explanation.

We can see this point by contrasting causal explanation with the explanation of one law by another, or with the 'preceding paper' relation I mentioned above. We need to sort the special van Fraassen–Duhem challenge we have been discussing from more general epistemological worries that make us question (as perhaps we always should) whether we really do have a good explanation. So let us introduce a fiction. God may tell you that Wollheim's paper is after mine, and that his paper is true. You have no dobuts about either of those propositions. This signifies nothing about the truth of my paper. Similarly, God tells you that Schroedinger's equation provides a completely satisfactory derivation of the phenomenological law of radioactive decay. You have no doubt that the derivation is correct. But you still have no reason to believe in Schroedinger's equation. On the other hand, if God tells you that the rotting of the roots is the cause of the yellowing of the leaves, or that the ionization produced by the negative charge explains the track in the cloud chamber, then you do have reason, conclusive reason, to believe that there is water in the tub and that there is an electron in the chamber.

3.   AN OBJECTION

I argue that inferences to the most likely cause have a different logical force than inferences to the best explanation. Larry Laudan raises a serious objection: 'It seems to me that your distinctions are plausible only because you insist (apparently arbitrarily) on countenancing a pragmatic view of theoretical laws and a non-pragmatic view of causal talk.'[5] In order to explain why I think this distinction is not arbitrary, I will lay out two very familiar views of explanation, one that underlies the deductive–nomological (D–N) model,[6] and the second, the view of Duhem. Van Fraassen challenges the realist to give an account of explanation that shows *why* the success of the explanation, coupled with the truth of the explanandum, argues for the truth of the explanans. I said there was an answer to this question in the case of causal inference. Similarly, I think there is an answer in the D–N account. If the D–N model is a correct account of what explanation is like, I agree that my distinction is arbitrary; but this is not so if Duhem is right.

If we could imagine that our explanatory laws *made* the phenomenological laws true, that would meet van Fraassen's challenge. But there is another, more plausible account that would do just as well. I discuss this account in great detail in the next essay. Adolf Grünbaum gives a brief sketch of the view:

It is crucial to realize that while (a more comprehensive law) *G* entails (a less comprehensive law) *L* logically, thereby providing an explanation of *L*, *G* is *not* the 'cause' of *L*. More specifically, laws are explained *not* by showing the regularities they affirm to be products of the operation of *causes* but rather by recognizing their truth to be special cases of more comprehensive truths.[7]

For any specific situation the fundamental laws are supposed to make the same claims as the more concrete phenomenological laws that they explain. This is borne out

---

[5] In correspondence, dated 15 September 1981.

[6] For a description of the deductive–nomological model of explanation, see C. G. Hempel, *Philosophy of Natural Science* (Englewood Cliffs, N.J.: Prentice-Hall, 1966).

[7] Adolf Grünbaum, 'Science and Ideology', *The Scientific Monthly* (July 1954), pp. 13–19, italics in the original.

by the fact that the phenomenological laws can be deduced from the fundamental laws, once a description of the situation is supplied. If the phenomenological laws have got it right, then so too do the fundamental, at least in that situation. There is still an inductive problem: are the fundamental laws making the right generalization across situations? But at least we see why the success of the explanation requires the truth of the explanans. To explain a phenomenological law is to restate it, but in a sufficiently abstract and general way that states a variety of other phenomenological laws as well. Explanatory laws are true statements of what happens; but unlike phenomenological laws, they are economical ways to say a lot.

This may seem straightforward. What else could explanation be? But contrast Duhem. Duhem believes that phenomena in nature fall roughly into natural kinds. The realist looks for something that unifies the members of the natural kind, something they all have in common; but Duhem denies that there is anything. There is nothing more than the rough facts of nature that sometimes some things behave like others, and what happens to one is a clue to what the others will do. Explanations provide a scheme that allows us to make use of these clues. Light and electricity behave in similar ways, but the procedures for drawing the analogies are intricate and difficult. It is easier for us to postulate the electromagnetic field and Maxwell's four laws, to see both light and electricity as a manifestation of one single underlying feature. There is no such feature, but if we are careful, we are better off to work with these fictional unifiers than to try to comprehend the vast array of analogies and disanalogies directly. The explanatory schemes we posit work as well as they do, even to producing novel productions, because phenomena do roughly fall into natural kinds. But in fact the phenomena are genuinely different. They only resemble each other some times in some ways, and the D-N attempt to produce one true description for all the members of the same class must inevitably fail. We cannot expect to find an explanatory law that will describe two phenomena that are in fact different, and yet will be true of both. What we can require of explanation is a scheme that allows us to exploit what similarities there are.

These are very cursory descriptions of the two views. But it is enough to see that the two embody quite different conceptions of explanation. Nor is it just a matter of choosing which to pursue, since they are joined to distinct metaphysical pictures. In practice the two conceptions meet; for in real life explanations, failure of deductivity is the norm. Duhem predicts this. But proponents of the D–N model can account for the practical facts as well. They attribute the failure of deductivity not to the lack of unity in nature, but to the failings of each particular theory we have to hand.

The difference between the two conceptions with respect to van Fraassen's challenge may be obscured by this practical convergence. We sometimes mistakenly assume that individual explanations, on either account, will look the same. Van Fraassen himself seems to suppose this; for he requires that the empirical substructures provided by a theory should be isomorphic to the true structures of the phenomena. But Duhem says that there can be at best a rough match. If Duhem is right, there will be no wealth of truly deductive explanations no matter how well developed a scientific discipline we look to.

Duhem sides with the thinkers who say 'A physical theory is an abstract system whose aim is to *summarize* and to classify *logically* a group of experimental laws without aiming to explain these laws', where 'to explain (explicate, *explicare*) is to strip reality of the appearances covering it like a veil, in order to see the bare reality itself.'[8] In an effort to remain metaphysically neutral, we might take an account of explanation which is more general than either Duhem's or the D–N story: to explain a collection of phenomenological laws is to give a physical theory of them, a physical theory in Duhem's sense, one that summarizes the laws and logically classifies them; only now we remain neutral as to whether we are also called upon to explain in the deeper sense of stripping away appearances. This is the general kind of account I have been supposing throughout this essay.

There is no doubt that we can explain in this sense.

[8] Pierre Duhem, *The Aim and Structure of Physical Theory*, op. cit., p. 7.

Physical theories abound, and we do not have to look to the future completion of science to argue that they are fairly successful at summarizing and organizing; that is what they patently do now. But this minimal, and non-question-begging, sense of explanation does not meet van Fraassen's challenge. There is nothing about successful organization that requires truth. The stripped down characterization will not do. We need the full paraphernalia of the D–N account to get the necessary connection between truth and explanation. But going beyond the stripped down view to the full metaphysics involved in a D–N account is just the issue in question.

There is still more to Laudan's criticism. Laudan himself has written a beautiful piece against inference to the best explanation.[9] The crux of his argument is this: it is a poor form of inference that repeatedly generates false conclusions. He remarks on case after case in the history of science where we now know our best explanations were false. Laudan argues that this problem plagues theoretical laws and theoretical entities equally. Of my view he says,

What I want to know, is what *epistemic* difference there is between the evidence we can have for a theoretical law (which you admit to be non-robust) and the evidence we can have for a theoretical entity— such that we are warranted in concluding that, say electrons and protons exist, but that we are not entitled to conclude that theoretical laws are probably true. It seems to me that the two are probably on an equal footing epistemically.

Laudan's favourite example is the electromagnetic aether, which 'had all sorts of independent sources of support for it collected over a century and a half'. He asks, 'Did the enviable successes of one- and two-fluid theories of electricity show that there really was an electrical fluid?'[10]

I have two remarks, the first very brief. Although the electromagnetic aether is one striking example, I think these cases are much rarer than Laundan does. So we have a historical dispute. The second remark bears on the first. I have been arguing that we must be committed to the existence of the cause if we are to accept a given causal account.

    [9] Larry Laudan, 'A Confutation of Convergent Realism'. *Philosophy of Science* 48 (March 1981), pp. 19–49.
    [10] In correspondence referred to in footnote 5.

The same is not true for counting a theoretical explanation good. The two claims get intertwined when we address the nontrivial and difficult question, when do we have reasonable grounds for counting a causal account acceptable? The fact that the causal hypotheses are part of a generally satisfactory explanatory theory is not enough, since success at organizing, predicting, and classifying is never an argument for truth. Here, as I have been stressing, the idea of direct experimental testing is crucial. Consider the example of the laser company, Spectra Physics, mentioned in the Introduction to these essays. Engineers at Spectra Physics construct their lasers with the aid of the quantum theory of radiation, non-linear optics, and the like; and they calculate their performance characteristics. But that will not satisfy their customers. To guarantee that they will get the effects they claim, they use up a quarter of a million dollars' worth of lasers every few months in test runs.

I think there is no general theory, other than Mill's methods, for what we are doing in experimental testing; we manipulate the cause and look to see if the effects change in the appropriate manner. For specific causal claims there are different detailed methodologies. Ian Hacking, in 'Experimentation and Scientific Realism', gives a long example of the use of Stanford's Peggy II to test for parity violations in weak neutral currents. There he makes a striking claim:

The experimentalist does not believe in electrons because, in the words retrieved from medieval science by Duhem, they 'save the phenomena'. On the contrary, we believe in them because we use them to *create* new phenomena, such as the phenomenon of parity violation in weak neutral current interactions.[11]

I agree with Hacking that when we can manipulate our theoretical entities in fine and detailed ways to intervene in other processes, then we have the best evidence possible for our claims about what they can and cannot do; and theoretical entities that have been warranted by well-tested causal claims like that are seldom discarded in the progress of science.

[11] See Ian Hacking, 'Experimentation and Scientific Realism', *Philosophical Topics* (forthcoming).

4. CONCLUSION

I believe in theoretical entities. But not in theoretical laws. Often when I have tried to explain my views on theoretical laws, I have met with a standard realist response: 'How *could* a law explain if it weren't true?' Van Fraassen and Duhem teach us to retort, 'How could it explain if it *were* true?' What is it about explanation that guarantees truth? I think there is no plausible answer to this question when one law explains another. But when we reason about theoretical entities the situation is different. The reasoning is causal, and to accept the explanation is to admit the cause. There is water in the barrel of my lemon tree, or I have no explanation for its ailment, and if there are no electrons in the cloud chamber, I do not know why the tracks are there.

For Phenomenological Laws

0. INTRODUCTION

A long tradition distinguishes fundamental from phenomenological laws, and favours the fundamental. Fundamental laws
are true in themselves; phenomenological laws hold only on
account of more fundamental ones. This view embodies an
extreme realism about the fundamental laws of basic explanatory theories. Not only are they true (or would be if we had
the right ones), but they are, in a sense, more true than the
phenomenological laws that they explain. I urge just the
reverse. I do so not because the fundamental laws are about
unobservable entities and processes, but rather because of
the nature of theoretical explanation itself. As I have often
urged in earlier essays, like Pierre Duhem, I think that the
basic laws and equations of our fundamental theories organize and classify our knowledge in an elegant and efficient
manner, a manner that allows us to make very precise calculations and predictions. The great explanatory and predictive
powers of our theories lies in their fundamental laws. Nevertheless the *content* of our scientific knowledge is expressed
in the phenomenological laws.

Suppose that some fundamental laws are used to explain
a phenomenological law. The ultra-realist thinks that the
phenomenological law is true *because of* the more fundamental laws. One elementary account of this is that the
fundamental laws make the phenomenological laws true.
The truth of the phenomenological laws derives from the
truth of the fundamental laws in a quite literal sense—
something like a causal relation exists between them. This is
the view of the seventeenth-century mechanical philosophy
of Robert Boyle and Robert Hooke. When God wrote the
Book of Nature, he inscribed the fundamental laws of
mechanics and he laid down the initial distribution of matter
in the universe. Whatever phenomenological laws would be
true fell out as a consequence. But this is not only the view of

the seventeenth-century mechanical philosophy. It is a view that lies at the heart of a lot of current-day philosophy of science—particularly certain kinds of reductionism—and I think it is in part responsible for the widespread appeal of the deductive-nomological model of explanation, though certainly it is not a view that the original proponents of the D-N model, such as Hempel, and Grünbaum, and Nagel, would ever have considered. I used to hold something like this view myself, and I used it in my classes to help students adopt the D-N model. I tried to explain the view with two stories of creation.

Imagine that God is about to write the Book of Nature with Saint Peter as his assistant. He might proceed in the way that the mechanical philosophy supposed. He himself decided what the fundamental laws of mechanics were to be and how matter was to be distributed in space. Then he left to Saint Peter the laborious but unimaginative task of calculating what phenomenological laws would evolve in such a universe. This is a story that gives content to the reductionist view that the laws of mechanics are fundamental and all the rest are epi-phenomenal.

On the other hand, God may have had a special concern for what regularities would obtain in nature. There were to be no distinctions among laws: God himself would dictate each and every one of them—not only the laws of mechanics, but also the laws of chemical bonding, of cell physiology, of small group interactions, and so on. In this second story Saint Peter's task is far more demanding. To Saint Peter was left the difficult and delicate job of finding some possible arrangement of matter at the start that would allow all the different laws to work together throughout history without inconsistency. On this account all the laws are true at once, and none are more fundamental than the rest.

The different roles of God and Saint Peter are essential here: they make sense of the idea that, among a whole collection of laws every one of which is supposed to be true, some are more basic or more true than others. For the seventeenth-century mechanical philosophy, God and the Book of Nature were legitimate devices for thinking of laws and the relations among them. But for most of us nowadays these stories are mere metaphors. For a long time I used the metaphors, and hunted for some non-metaphorical analyses. I now think that it cannot be done. Without God and the Book of Nature there is no sense to be made of the idea that one law derives from another in nature, that the fundamental laws are basic and that the others hold literally 'on account of' the fundamental ones.

Here the D-N model of explanation might seem to help. In order to explain and defend our reductionist views, we look for some quasi-causal relations among laws in nature. When we fail to find any reasonable way to characterize these relations in nature, we transfer our attention to language. The deductive relations that are supposed to hold between the laws of a scientific explanation act as a formal-mode stand-in for the causal relations we fail to find in the material world. But the D-N model itself is no argument for realism once we have stripped away the questionable metaphysics. So long as we think that deductive relations among statements of law mirror the order of responsibility among the laws themselves, we can see why explanatory success should argue for the truth of the explaining laws. Without the metaphysics, the fact that a handful of elegant equations can organize a lot of complex information about a host of phenomenological laws is no argument for the truth of those equations. As I urged in the last essay, we need some story about what the connection between the fundamental equations and the more complicated laws is supposed to be. There we saw that Adolf Grünbaum has outlined such a story. His outline I think coincides with the views of many contemporary realists. Grünbaum's view eschews metaphysics and should be acceptable to any modern empiricist. Recall that Grünbaum says:

It is crucial to realize that while (a more comprehensive law) *G* entails (a less comprehensive law) *L* logically, thereby providing an explanation of *L*, *G* is not the 'cause' of *L*. More specifically, laws are explained *not* by showing the regularities they affirm to be products of the operation of *causes* but rather by recognizing their truth to be special cases of more comprehensive truths.[1]

[1] A. Grünbaum, 'Science and Ideology', *The Scientific Monthly* (July 1954), p. 14, italics in original.

I call this kind of account of the relationship between fundamental and phenomenological laws a *generic-specific* account. It holds that in any particular set of circumstances the fundamental explanatory laws and the phenomenological laws that they explain both make the same claims. Phenomenological laws are what the fundamental laws *amount* to in the circumstances at hand. But the fundamental laws are superior because they state the facts in a more general way so as to make claims about a variety of different circumstances as well.

The generic-specific account is nicely supported by the deductive-nomological model of explanation: when fundamental laws explain a phenomenological law, the phenomenological law is deduced from the more fundamental in conjunction with a description of the circumstances in which the phenomenological law obtains. The deduction shows just what claims the fundamental laws make in the circumstances described.

But explanations are seldom in fact deductive, so the generic-specific account gains little support from actual explanatory practice. Wesley Salmon[2] and Richard Jeffrey,[3] and now many others, have argued persuasively that explanations are not arguments. But their views seem to bear most directly on the explanations of single events; and many philosophers still expect that the kind of explanations that we are concerned with here, where one law is derived from others more fundamental, will still follow the D-N form. One reason that the D-N account often seems adequate for these cases is that it starts looking at explanations only after a lot of scientific work has already been done. It ignores the fact that explanations in physics generally begin with a model. The calculation of the small signals properties of amplifiers, which I discuss in the next section, is an example.[4] We first

[2] See Wesley Salmon, 'Statistical Explanation', in Wesley Salmon (ed.), *Statistical Explanation and Statistical Relevance* (Pittsburgh: University of Pittsburgh Press, 1971).

[3] See R. C. Jeffrey, 'Statistical Explanation vs. Statistical Inference', in Wesley Salmon, op. cit.

[4] This example is taken from my joint paper with Jon Nordby, 'How Approximations Take Us Away from Theory and Towards the Truth' (unpublished manuscript: Stanford University and Pacific Lutheran University).

decide which model to use—perhaps the T-model, perhaps the hybrid-π model. Only then can we write down the equations with which we will begin our derivation. Which model is the right one? Each has certain advantages and disadvantages. The T-model approach to calculating the midband properties of the CE stage is direct and simple, but if we need to know how the CE stage changes when the bias conditions change, we would need to know how all the parameters in the transistor circuit vary with bias. It is incredibly difficult to produce these results in a T-model. The T-model also lacks generality, for it requires a new analysis for each change in configuration, whereas the hybrid-π model approach is most useful in the systematic analysis of networks. This is generally the situation when we have to bring theory to bear on a real physical system like an amplifier. For different purposes, different models with different incompatible laws are best, and there is no single model which just suits the circumstances. The facts of the situation do not pick out one right model to use.

I will discuss models at length in the next few essays. Here I want to lay aside my worries about models, and think about how derivations proceed once a model has been chosen. Proponents of the D-N view tend to think that at least then the generic-specific account holds good. But this view is patently mistaken when one looks at real derivations in physics or engineering. It is never strict deduction that takes you from the fundamental equations at the beginning to the phenomenological laws at the end. Instead we require a variety of different approximations. In any field of physics there are at most a handful of rigorous solutions, and those usually for highly artificial situations. Engineering is worse.

Proponents of the generic-specific account are apt to think that the use of approximations is no real objection to their view. They have a story to tell about approximations: the process of deriving an approximate solution parallels a D-N explanation. One begins with some general equations which we hold to be exact, and a description of the situation to which they apply. Often it is difficult, if not impossible, to solve these equations rigorously, so we rely on our description of the situation to suggest approximating procedures.

In these cases the approximate solution is just a stand-in. We suppose that the rigorous solution gives the better results; but because of the calculational difficulties, we must satisfy ourselves with some approximation to it.

Sometimes this story is not far off. Approximations occasionally work just like this. Consider, for example, the equation used to determine the equivalent air speed, V_E, of a plane (where P_T is total pressure, P_0 is ambient pressure, ρ_S is sea level density, and M is a Mach number):

$$V_E = \left[2\left(\frac{P_T - P_0}{\rho_S}\right) \times \left(\frac{1}{1 + M^2/4 + M^4/40}\right) \right]^{\frac{1}{2}} \quad (6.1)$$

The value for V_E determined by this equation is close to the plane's true speed.

Approximations occur in connection with (6.1) in two distinct but typical ways. First, the second term,

$$\frac{1}{1 + M^2/4 + M^4/40}$$

is significant only as the speed of the plane approaches Mach One. If $M < 0.5$, the second term is discarded because the result for V_E given by

$$\left[2\left(\frac{P_T - P_0}{\rho_S}\right) \right]^{\frac{1}{2}}$$

will differ insignificantly from the result given by (6.1). Given this insignificant variation, we can approximate V_E by using

$$V_E = \left[2\left(\frac{P_T - P_0}{\rho_S}\right) \right]^{\frac{1}{2}} \quad (6.2)$$

for $M < 0.5$. Secondly, (6.1) is already an approximation, and not an exact equation. The term

$$\frac{1}{1 + M^2/4 + M^4/40}$$

has other terms in the denominator. It is a Taylor series expansion. The next term in the expansion is $M^6/1600$, so we get

$$\frac{1}{1 + M^2/4 + M^4/40 + M^6/1600},$$

in the denominator, and so on. For Mach numbers less than one, the error that results from ignoring this third term is less than one per cent, so we truncate and use only two terms.

Why does the plane travel with a velocity, V, roughly equal to

$$\left[2\left(\frac{P_T - P_0}{\rho_S}\right)\right]^{\frac{1}{2}}?$$

Because of equation (6.1). In fact the plane is really travelling at a speed equal to

$$\left[2\left(\frac{P_T - P_0}{\rho_S}\right) \times \frac{1}{1 + M^2/4 + M^4/40 + \dots}\right]^{\frac{1}{2}}.$$

But since M is less than 0.5, we do not notice the difference. Here the derivation of the plane's speed parallels a covering law account. We assume that equation (6.1) is a true law that covers the situation the plane is in. Each step away from equation (6.1) takes us a little further from the true speed. But each step we take is justified by the facts of the case, and if we are careful, we will not go too far wrong. The final result will be close enough to the true answer.

This is a neat picture, but it is not all that typical. Most cases abound with problems for the generic-specific account. Two seem to me especially damaging: (1) practical approximations usually improve on the accuracy of our fundamental laws. Generally the doctored results are far more accurate than the rigorous outcomes which are strictly implied by the laws with which we begin. (2) On the generic-specific account the steps of the derivation are supposed to show how the fundamental laws make the same claims as the phenomenological laws, given the facts of the situation. But seldom are the facts enough to justify the derivation. Where approximations are called for, even a complete knowledge of the circumstances may not provide the additional premises necessary to deduce the phenomenological laws from the fundamental equations that explain them. Choices must be made which are not dictated by the facts. I have already mentioned that this is so with the choice of models. But it is also the case with approximation procedures: the choice is constrained, but not dictated by the facts, and different choices give rise to different, incompatible results. The generic-specific account fails because the content of the phenomenological laws we derive is not contained in the fundamental laws which explain them.

These two problems are taken up in turn in the next two sections. A lot of the argumentation, especially in Section 2, is taken from a paper written jointly by Jon Nordby and me, 'How Approximations Take Us Away from Theory and Towards the Truth'.[5] This paper also owes a debt to Nordby's 'Two Kinds of Approximations in the Practice of Science'.[6]

1.   APPROXIMATIONS THAT IMPROVE ON LAWS

On the generic-specific account, any approximation detracts from truth. But it is hard to find examples of this at the level where approximations connect theory with reality, and that is where the generic-specific account must work if it is to ensure that the fundamental laws are true in the real world. Generally at this level approximations take us away from theory and each step away from theory moves closer towards the truth. I illustrate with two examples from the joint paper with Jon Nordby.

1.1.   *An Amplifier Model*

Consider an amplifier constructed according to Figure 6.1. As I mentioned earlier, there are two ways to calculate the small signal properties of this amplifier, the T-model and the hybrid-π model. The first substitutes a circuit model for the transistor and analyses the resulting network. The second characterizes the transistor as a set of two-port parameters and calculates the small signal properties of the amplifier

---

[5] N. Cartwright and J. Nordby, 'How Approximations Take Us Away from Theory and Towards the Truth' (unpublished manuscript: Stanford University and Pacific Lutheran University).

[6] See Jon Nordby, 'Two Kinds of Approximation in the Practice of Science' (unpublished manuscript: Pacific Lutheran University).

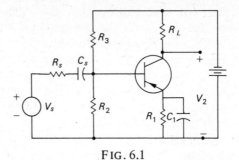

FIG. 6.1

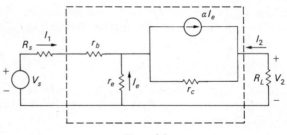

FIG. 6.2

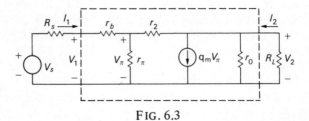

FIG. 6.3

in terms of these parameters. The two models are shown in Figures 6.2 and 6.3 inside the dotted area.

The application of these transistor models in specific situations gives a first rough approximation of transistor parameters at low frequencies. These parameters can be theoretically estimated without having to make any measurements on the actual circuit. But the theoretical estimates are often grossly inaccurate due to specific causal features present in the actual circuit, but missing from the models.

One can imagine handling this problem by constructing a larger, more complex model that includes the missing causal features. But such a model would have to be highly specific to the circuit in question and would thus have no general applicability.

Instead a different procedure is followed. Measurements of the relevant parameters are made on the actual circuit under study, and then the measured values rather than the theoretically predicted values are used for further calculations in the original models. To illustrate, consider an actual amplifier, built and tested such that $I_E = 1$ ma; $R_L = 2.7 \| 15 = 2.3$ k ohm; $R_1 = 1$ k ohm. $\beta = 162$ and $R_S = 1$ k ohm and assume $r_b = 50$ ohms. The theoretical expectation of midband gain is

$$|A_V| = \frac{R_L}{r_e + (r_b + R_S)(1 - \alpha)} = \frac{2.3 \text{ k ohms}}{32 \text{ k ohms}} = 72. \quad (6.3)$$

The actual measured midband gain for this amplifier with an output voltage at $f = 2$ kHz and source voltage at 1.8 mv and 2 kHz, is

$$A_V, \text{meas} = \frac{80 \text{ mv}}{1.8 \text{ mv}} = 44. \quad (6.4)$$

This result is not even close to what theory predicts. This fact is explained causally by considering two features of the situation: first, the inaccuracy of the transistor model due to some undiagnosed combination of causal factors; and second, a specifically diagnosed omission—the omission of equivalent series resistance in the bypass capacitor. The first inaccuracy involves the value assigned to r_e by theory. In theory, $r_e = kT/qI_E$, which is approximately $25.9/I_E$. The actual measurements indicate that the constant of proportionality for this type of transistor is 30 mv, so $r_e = 30/I_E$, not $25.9/I_E$.

Secondly, the series resistance is omitted from the ideal capacitor. But real electrolytic capacitors are not ideal. There is leakage of current in the electrolyte, and this can be modelled by a resistance in series with the capacitor. This series resistance is often between 1 and 10 ohms, sometimes as high as 25 ohms. It is fairly constant at low frequency but increases with increased frequency and also with increased temperature. In this specific case, the series resistance, r_{C_1}, is measured as 12 ohms.

We must now modify (6.3) to account for these features, given our measured result in (6.4) of A_V, meas. = 44:

$$|A_V| = \frac{R_L}{r_e + (r_b + R_S)(1 - \alpha) + r_{C_1}}. \tag{6.5}$$

Solving this equation gives us a predicted midband gain of 47.5, which is sufficiently close to the measured midband gain for most purposes.

Let us now look back to see that this procedure is very different from what the generic-specific account supposes. We start with a general abstract equation, (6.3); make some approximations; and end with (6.5), which gives rise to detailed phenomenological predictions. Thus superficially it may look like a D-N explanation of the facts predicted. But unlike the covering laws of D-N explanations, (6.3) as it stands is not an equation that really describes the circuits to which it is applied. (6.3) is refined by accounting for the specific causal features of each individual situation to form an equation like (6.5). (6.5) gives rise to accurate predictions, whereas a rigorous solution to (6.3) would be dramatically mistaken.

But one might object: isn't the circuit model, with no resistance added in, just an idealization? And what harm is that? I agree that the circuit model is a very good example of one thing we typically mean by the term *idealization*. But how can that help the defender of fundamental laws? Most philosophers have made their peace with idealizations: after all, we have been using them in mathematical physics for well over two thousand years. Aristotle in proving that the rainbow is no greater than a semi-circle in *Meterologica* III. 5 not only treats the sun as a point, but in a blatant falsehood puts the sun and the reflecting medium (and hence the rainbow itself) the same distance from the observer. Today we still make the same kinds of idealizations in our celestial theories.[7] Nevertheless, we have managed to discover

---

[7] See Hilary Putnam, 'The "Corroboration" of Theories', *Philosophical Papers*, Vol. 1 (Cambridge: Cambridge University Press, 1975) for a nice discussion of this.

the planet Neptune, and to keep our satellites in space. Idealizations are no threat to the progress of science.

But what solace is this to the realist? How do idealizations save the truth of the fundamental laws? The idea seems to be this. To call a model an idealization is to suggest that the model is a simplification of what occurs in reality, usually a simplification which omits some relevant features, such as the extended mass of the planets or, in the example of the circuit model, the resistance in the bypass capacitor. Sometimes the omitted factors make only an insignificant contribution to the effect under study. But that does not seem to be essential to idealizations, especially to the idealizations that in the end are applied by engineers to study real things. In calling something an idealization it seems not so important that the contributions from omitted factors be small, but that they be ones for which we know how to correct. If the idealization is to be of use, when the time comes to apply it to a real system we had better know how to add back the contributions of the factors that have been left out. In that case the use of idealizations does not seem to counter realism: either the omitted factors do not matter much, or in principle we know how to treat them.

In the sense I just described, the circuit model is patently an idealization. We begin with equation (6.3), which is inadequate; we know the account can be improved—Nordby and I show how. But the improvements come at the wrong place for the defender of fundamental laws. They come from the ground up, so-to-speak, and not from the top down. We do not modify the treatment by deriving from our theoretical principles a new starting equation to replace (6.3). It is clear that we could not do so, since only part of the fault is diagnosed. What we do instead is to add a phenomenological correction factor, a factor that helps produce a correct description, but that is not dictated by fundamental law.

But could we not 'in principle' make the corrections right at the start, and write down a more accurate equation from the beginning? That is just the assumption I challenge. Even if we could, why do we think that by going further and further backwards, trying to get an equation that will be right when all the significant factors are included, we will eventually get something simple which looks like one of the fundamental laws of our basic theories? Recall the discussion of cross-effects from Essay 3. There I urged that we usually do not have any uniform procedure for 'adding' interactions. When we try to write down the 'more correct' equations, we get a longer and longer list of complicated laws of different forms, and not the handful of simple equations which could be fundamental in a physical theory.

Generality and simplicity are the substance of explanation. But they are also crucial to application. In engineering, one wants laws with a reasonably wide scope, models that can be used first in one place then another. If I am right, a law that actually covered any specific case, without much change or correction, would be so specific that it would not be likely to work anywhere else. Recall Bertrand Russell's objection to the 'same cause, same effect' principle:

The principle 'same cause, same effect,' which philosophers imagine to be vital to science, is therefore utterly otiose. As soon as the antecedents have been given sufficiently fully to enable the consequents to be calculated with some exactitude, the antecedents have become so complicated that it is very unlikely they will ever recur. Hence, if this were the principle involved, science would remain utterly sterile.[8]

Russell's solution is to move to functional laws which state relations between properties (rather than relations between individuals). But the move does not work if we want to treat real, complex situations with precision. Engineers are comfortable with functions. Still they do not seem able to find functional laws that allow them to calculate consequences 'with some exactitude' and yet are not 'so complicated that it is very unlikely they will ever recur'. To find simple laws that we can use again and again, it looks as if we had better settle for laws that patently need improvement. Following Russell, it seems that if we model approximation on D-N explanation, engineering 'would remain utterly sterile'.

[8] Bertrand Russell, 'On the Notion of Cause with Application to the Problem of Free Will', in H. Feigl and M. Brodbeck (eds), *Readings in Philosophy of Science* (New York: Appleton-Century-Crofts, 1953), p. 392.

1.2. *Exponential Decay*

The second example concerns the derivation of the exponential decay law in quantum mechanics. I will describe this derivation in detail, but the point I want to stress can be summarized by quoting one of the best standard texts, by Eugen Merzbacher: 'The fact remains that the exponential decay law, for which we have so much empirical support in radioactive processes, is not a rigorous consequence of quantum mechanics but the result of somewhat delicate approximations.'[9]

The exponential decay law is a simple, probabilistically elegant law, for which—as Merzbacher says—we have a wealth of experimental support. Yet it cannot be derived exactly in the quantum theory. The exponential law can only be derived by making some significant approximation. In the conventional treatment the rigorous solution is not pure exponential, but includes several additional terms. These additional terms are supposed to be small, and the difference between the rigorous and the approximate solution will be unobservable for any realistic time periods. The fact remains that the data, together with any reasonable criterion of simplicity (and some such criterion must be assumed if we are to generalize from data to laws at all) speak for the truth of an exponential law; but such a law cannot be derived rigorously. Thus it seems that the approximations we make in the derivation take us closer to, not further from, the truth.

There are two standard treatments of exponential decay: the Weisskopf–Wigner treatment, which was developed in their classic paper of 1930,[10] and the more recent Markov treatment, which sees the exponential decay of an excited atom as a special case in the quantum theory of damping. We will look at the more recent treatment first. Here we consider an abstract system weakly coupled to a reservoir. The aim is to derive a general master equation for the

---

[9] Eugen Merzbacher, *Quantum Mechanics* (New York: John Wiley & Sons, 1970), pp. 484–5.

[10] V. Weisskopf and E. Wigner, 'Die Rechnung der natürlichen Linienbreite auf Grund der Diracschen Lichttheorie', *Zeitschrift für Physik* 63 (1930), pp. 54–73.

evolution of the system. This equation is similar to the evolution equations of classical statistical mechanics. For the specific case in which we are interested, where the system is an excited atom and the reservoir the electromagnetic field, the master equation turns into the Pauli rate equation, which is the analogue of the exponential law when re-excitement may occur:

Pauli equation:

$$\frac{\partial S_i}{\partial t} = -\Gamma_i S_i + \sum_{k \neq i} \omega_{ik} S_k.$$

(Here S_j is the occupation probability of the jth state; Γ_j, the inverse of the lifetime; and ω_{jk} is the transition probability from state k to state j.)

The derivation of the master equation is quite involved. I will focus on the critical feature from my point of view — the Markov approximation. Generally such a derivation begins with the standard second-order perturbation expansion for the state x of the composite, system and reservoir, which in the interaction picture looks like this:

$$\chi(t) = \chi(t_0) + \frac{1}{i\hbar} \int_{t_0}^{t} [V(t' - t_0), \chi(t_0)] \, dt' +$$

$$+ \left(\frac{1}{i\hbar}\right)^2 \int_0^t dt' \int_0^{t'} dt'' [V(t' - t_0), [V(t'' - t_0), \chi(t_0)]].$$

Notice that the state of the system and reservoir at t depends on its entire past history through the integrals on the right-hand side of this equation. The point of the Markov approximation is to derive a differential equation for the state of the system alone such that the change in this state at a time depends only on the facts about the system at *that* time, and not on its past history. This is typically accomplished by two moves: (i) extending the time integrals which involve only reservoir correlations to infinity, on the grounds that the correlations in the reservoir are significant for only a short period compared to the periods over which we are observing the system; and (ii) letting $t - t_0 \to 0$, on the grounds that the periods of time considered for the system are small compared to its lifetime. The consequence is a master equation with the desired feature. As W. H. Louisell remarks in his chapter on damping:

We note that the r.h.s. [right hand side] of [the master equation] no longer contains time integrals over $S(t')$ [S is the state of the system alone] for times earlier than the present so that the future is now indeed determined by the present. We have assumed that the reservoir correlation times are zero on a time scale in which the system loses an appreciable amount of its energy . . . One sometimes refers to the Markoff approximation as a coarse-grained averaging.[11]

Thus the Markov approximation gives rise to the master equation; for an atom in interaction with the electromagnetic field, the master equation specializes to the Pauli equation; and the Pauli equation predicts exponential decay for the atom. Without the Markov approximation, the decay can at best be near exponential.

Let us now look at the Weisskopf–Wigner method, as it is employed nowadays. We begin with the exact Schroedinger equations for the amplitudes, but assume that the only significant coupling is between the excited and the de-excited states:

$$\frac{dc_e}{dt} = \sum_f g_{ef}^2 \int_0^t dt' \exp\{i(\omega_{eg} - \omega_f)(t - t')\}c_e(t')$$

(ω_{eg} is $E_e - E_g/\hbar$, for E_e the energy of the excited state, E_g the energy of the de-excited; ω_f is the frequency of the fth mode of the field; and g_{ef} is the coupling constant between the excited state and the fth mode. c_e is the amplitude in the excited state, no photons present.)

The first approximation notes that the modes of the field available to the de-exciting atom form a near continuum. (We will learn more about this in the next section.) So the sum over f can be replaced by an integral, to give

$$\frac{dc_e}{dt} = \int_{-\infty}^{+\infty} -g^2(\omega)\mathcal{D}(\omega)d\omega \int_0^t dt' \exp\{i(\omega_{eg} - \omega)(t - t')\}c_e(t').$$

We could now pull out the terms which are slowly varying in ω, and do the ω integral, giving a delta function in t:

[11] William H. Louisell, *Quantum Statistical Properties of Radiation* (New York: John Wiley & Sons, 1973), p. 341.

$$\frac{dc_e}{dt} = g^2(\omega_{eg})\mathcal{D}(\omega_{eg}) \int_0^t dt' c_e(t') \times$$

$$\times \int_{-\infty}^{+\infty} \exp\{i(\omega_{eg} - \omega)(t - t')\}d\omega$$

$$= g^2(\omega_{eg})\mathcal{D}(\omega_{eg})\pi \int_0^t dt' c_e(t')\delta(t - t'),$$

or setting $\gamma \equiv 2\pi g^2(\omega_{eg})\mathcal{D}(\omega_{eg})$,

$$\frac{dc_e}{dt} = -\frac{\gamma}{2} c_e(t)$$

and finally,

$$c_e(t) = \exp(-\gamma/2t).$$

But in moving so quickly we lose the Lamb shift—a small displacement in the energy levels discovered by Willis Lamb and R. C. Retherford in 1947. It will pay to do the integrals in the opposite order. In this case, we note that $c_e(t)$ is itself slowly varying compared to the rapid oscillations from the exponential, and so it can be factored out of the t' integral, and the upper limit of that integral can be extended to infinity. Notice that the extension of the t' limit is very similar to the Markov approximation already described, and the rationale is similar. We get

$$\frac{dc_e}{dt} = \int_{-\infty}^{+\infty} g^2(\omega)\mathcal{D}(\omega)c_e(t)\exp\{i(\omega_{eg} - \omega)t\}d\omega \times$$

$$\times \int_0^\infty \exp\{-i(\omega_{eg} - \omega)t'\}dt'$$

$$= \int_{-\infty}^{+\infty} g^2(\omega)\mathcal{D}(\omega)c_e(t)\exp\{i(\omega_{eg} - \omega)t\}\,d\omega \times$$

$$\times \left\{\pi\delta(\omega_{eg} - \omega) + i\mathscr{P}\left(\frac{1}{\omega_{eg} - \omega}\right)\right\},$$

or, setting $\gamma \equiv 2\pi g^2(\omega_{eg})\mathcal{D}(\omega_{eg})$ and

$$\Delta\omega \equiv -\int_{-\infty}^{+\infty} \frac{g^2(\omega)\mathcal{D}(\omega)}{\omega_{eg} - \omega}\,d\omega$$

$(\mathscr{P}(x) = $ principal part of x.)

$$\frac{dc_e}{dt} = -\left(\frac{\gamma}{2} + i\Delta\omega\right)c_e(t).$$

Here $\Delta\omega$ is the Lamb shift. The second method, which results in a Lamb shift as well as the line-broadening γ, is what is usually now called 'the Weisskopf–Wigner' method.

We can try to be more formal and avoid approximation altogether. The obvious way to proceed is to evaluate the Laplace transform, which turns out to be

$$\mathbf{c}_e(\delta) \equiv \int_0^\infty \exp(-\delta t)c_e(t)dt = \cfrac{1}{\delta + i \sum_f \left(\cfrac{g_{ef}^2}{\omega_{eg} - \omega_f + i\delta}\right)}$$

then

$$\mathbf{c}_e(t) = \frac{1}{2\pi i} \int_{\epsilon - i\infty}^{\epsilon + i\infty} \exp(\delta t)\mathbf{c}_e(\delta)d\delta.$$

To solve this equation, the integrand must be defined on the first and second Riemann sheets. The method is described clearly in Goldberger and Watson's text on collision theory.[12]

The primary contribution will come from a simple pole $\Delta\epsilon$ such that

$$\frac{i}{\hbar\Delta\epsilon} \equiv \frac{\gamma}{2} + i\Delta\omega = \lim_{\delta \to 0^+} \sum_f \frac{g_{ef}}{\omega_{eg} - \omega_f + i\delta}.$$

This term will give us the exponential we want:

$$c_e(t) = \exp\left\{-\left(\frac{\gamma}{2} + i\Delta\omega\right)t\right\}.$$

But this is not an exact solution. As we distort the contour on the Riemann sheets, we cross other poles which we have not yet considered. We have also neglected the integral around the final contour itself. Goldberger and Watson calculate that this last integral contributes a term proportional to $\hbar^{3/2}/t^{3/2}E_e$. They expect that the other poles will add only

[12] See Marvin L. Goldberger and Kenneth M. Watson, *Collision Theory* (New York: John Wiley & Sons, 1964), Chapter 8.

negligible contributions as well, so that the exact answer will be a close approximation to the exponential law we are seeking; a close approximation, but still only an approximation. If a pure exponential law is to be derived, we had better take our approximations as improvements on the initial Schroedinger equation, and not departures from the truth.

Is there no experimental test that tells which is right? Is decay really exponential, or is the theory correct in predicting departures from the exponential law when t gets large? There was a rash of tests of the experimental decay law in the middle 1970s, spurred by E. T. Jaynes's 'neo-classical' theory of the interaction of matter with the electromagnetic field, a theory that did surprisingly well at treating what before had been thought to be pure quantum phenomena. But these tests were primarily concerned with Jaynes's claim that decay rates would depend on the occupation level of the initial state. The experiments had no bearing on the question of very long time decay behaviour. This is indeed a very difficult question to test. Rolf Winter[13] has experimented on the decay of Mn^{56} up to 34 half-lives, and D. K. Butt and A. R. Wilson on the alpha decay of radon for over 40 half-lives.[14] But, as Winter remarks, these lengths of time, which for ordinary purposes are quite long, are not relevant to the differences I have been discussing, since 'for the radioactive decay of Mn^{56} . . . non-exponential effects should not occur before roughly 200 half-lives. In this example, as with all the usual radioactive decay materials, nothing observable should be left long before the end of the exponential region'.[15] In short, as we read in a 1977 review article by A. Pais, 'experimental situations in which such deviations play a role have not been found to date.'[16] The times before the differences emerge are just too long.

[13] See Rolf Winter, 'Large-Time Exponential Decay and "Hidden Variables" ', *Physical Review* 126 (1962), pp. 1152–3.

[14] See D. K. Butt and A. R. Wilson, 'A Study of the Radioactive Decay Law', *Journal of Physics A: General Physics* 5 (1972), pp. 1248–51.

[15] Rolf Winter, op. cit., p. 1152.

[16] A. Pais, 'Radioactivity's Two Early Puzzles', *Reviews of Modern Physics* 49 (1977), p. 936.

2. APPROXIMATIONS NOT DICTATED
BY THE FACTS

Again, I will illustrate with two examples. The examples show how the correct approximation procedure can be un-determined by the facts. Both are cases in which the very same procedure, justified by exactly the same factual claims, gives different results depending on when we apply it: the same approximation applied at different points in the deriva-tion yields two different incompatible predictions. I think this is typical of derivation throughout physics; but in order to avoid spending too much space laying out the details of different cases, I will illustrate with two related phenomena: (*a*) the Lamb shift in the *excited* state of a single two-level atom; and (*b*) the Lamb shift in the *ground* state of the atom.

2.1. *The Lamb Shift in the Excited State*

Consider again spontaneous emission from a two-level atom. The traditional way of treating exponential decay derives from the classic paper of V. Weisskopf and Eugene Wigner in 1930, which I described in the previous section. In its present form the Weisskopf–Wigner method makes three important approximations: (1) the rotating wave approximation; (2) the replacement of a sum by an integral over the modes of the electromagnetic field and the factoring out of terms that vary slowly in the frequency; and (3) factoring out a slowly-varying term from the time integral and extending the limit on the integral to infinity. I will discuss the first approximation below when we come to consider the level shift in the ground state. The second and third are familiar from the last section. Here I want to concentrate on how they affect the Lamb shift in the excited state.

Both approximations are justified by appealing to the physical characteristics of the atom-field pair. The second approximation is reasonable because the modes of the field are supposed to form a near continuum; that is, there is a very large number of very closely spaced modes. This allows us to replace the sum by an integral. The integral is over a product of the coupling constant as a function of the fre-quency, ω, and a term of the form $\exp(-i\omega t)$. The coupling

constant depends on the interaction potential for the atom
and the field, and it is supposed to be relatively constant in
ω compared to the rapidly oscillating exponential. Hence
it can be factored outside the integral with little loss of
accuracy. The third approximation is similarly justified by
the circumstances.

What is important is that, although each procedure is
separately rationalized by appealing to facts about the atom
and the field, it makes a difference in what order they are
applied. This is just what we saw in the last section. If we
begin with the third approximation, and perform the t
integral *before* we use the second approximation to evaluate
the sum over the modes, we predict a Lamb shift in the
excited state. If we do the approximations and take the
integrals in the reverse order—which is essentially what
Weisskopf and Wigner did in their original paper—we lose
the Lamb shift. The facts that we cite justify both procedures, but the facts do not tell us in what order to apply
them. There is nothing about the physical situation that
indicates which order is right other than the fact to be
derived: a Lamb shift is observed, so we had best do first (3),
then (2). Given all the facts with which we start about the
atom and the field and about their interactions, the Lamb
shift for the excited state fits the fundamental quantum
equations. But we do not derive it from them.

One may object that we are not really using the same
approximation in different orders; for, applied at different
points, the same technique does not produce the same
approximation. True, the coefficients in t and in ω are slowly
varying, and factoring them out of the integrals results in
only a small error. But the exact size of the error depends on
the order in which the integrations are taken. The order that
takes first the t integral and then the ω integral is clearly
preferable, because it reduces the error.

Two remarks are to be made about this objection. Both
have to do with how approximations work in practice. First,
in this case it looks practicable to try to calculate and to
compare the amounts of error introduced by the two
approximations. But often it is practically impossible to
decide which of two procedures will lead to more accurate

results. For instance, we often justify dropping terms from an equation by showing that the coefficients of the omitted terms are small compared to those of the terms we retain. But as the next example will show, knowing the relative sizes of terms in the equation is not a sure guide to the exact effects in the solution, particularly when the approximation is embedded in a series of other approximations. This is just one simple case. The problem is widespread. As I argued in Essay 4, proliferation of treatments is the norm in physics, and very often nobody knows exactly how they compare. When the situation becomes bad enough, whole books may be devoted to sorting it out. Here is just one example, *The Theory of Charge Exchange* by Robert Mapleton. The primary purpose of the book is to explain approximating methods for cross sections and probabilities for electronic capture. But its secondary purpose is

to compare different approximate predictions with each other and with experimentally determined values. These comparisons should enable us to determine which approximating procedures are most successful in predicting cross sections for different ranges . . . they also should indicate which methods show most promise for additional improvement.[17]

Comparing approximations is often no easy matter.

Secondly, the objection assumes the principle 'the more accuracy, the better'. But this is frequently not so, for a variety of well-known reasons: the initial problem is set only to a given level of accuracy, and any accuracy in the conclusion beyond this level is spurious; or the use of certain mathematical devices, such as complex numbers, will generate excess terms which we do not expect to have any physical significance; and so on. The lesson is this: a finer approximation provides a better account than a rougher one only if the rougher approximation is not good enough. In the case at hand, the finer approximation is now seen to be preferable, not because it produces a quantitatively slightly more accurate treatment, but rather because it *exposes a qualitatively significant new phenomenon*—the Lamb shift in the ground state. If you look back at the equations in the last section

[17] Robert Mapleton, *The Theory of Charge Exchange* (New York: John Wiley & Sons, 1972), p. 1.

it is obvious that the first-ω-then t order misses an imaginary term: the amplitude to remain in the excited state has the form $\exp(-\frac{1}{2}\Gamma t)$ rather than $\exp\{-(\frac{1}{2}\Gamma + i\omega)t\}$. This additional imaginary term, $i\omega$, appears when the integrals are done in the reverse order, and it is this term that represents the Lamb shift. But what difference does this term make? In the most immediate application, for calculating decay probabilities, it is *completely* irrelevant, for the probability is obtained by multiplying together the amplitude $\exp\{-(\frac{1}{2}\Gamma + i\omega)t\}$, and its complex conjugate $\exp\{-(\frac{1}{2}\Gamma - i\omega)t\}$, in which case the imaginary part disappears and we are left with the well-known probability for exponential decay, $\exp(-\Gamma t)$.

The point is borne out historically. The first-ω-then-t order, which loses the Lamb shift, is an equivalent approximation to the *ansatz* which Weisskopf and Wigner use in their paper of 1930, and it was the absolutely conventional treatment for seventeen years. To calculate the value of the missing imaginary terms, one has to come face to face with divergences that arise from the Dirac theory of the electron, and which are now so notorious in quantum electrodynamics. These problems were just pushed aside until the remarkable experiments in 1947 by Willis Lamb and his student R. C. Retherford, for which Lamb later won the Nobel prize.

The Dirac theory, taking the spin of the electron into account, predicted an exact coincidence of the $2^2P_{1/2}$ and the $2^2S_{1/2}$ levels. There was a suspicion that this prediction was wrong. New microwave techniques developed during the war showed Lamb a way to find out, using the metastable $2^2S_{1/2}$ state of hydrogen. In the 1947 experiment the Lamb shift was discovered and within a month Bethe had figured a way to deal with the divergences. After the discovery of the Lamb shift the original Weisskopf–Wigner method had to be amended. Now we are careful to take the integrals in the first-t-then-ω order. But look at what Bethe himself has to say:

By very beautiful experiments, Lamb and Retherford have shown that the fine structure of the second quantum state of hydrogen does not agree with the prediction of the Dirac theory. The 2s level, which according to Dirac's theory should coincide with the $2p_{1/2}$ level, is actually higher than the latter by an amount of about 0.033 cm^{-1} or 1000 megacycles . . .

Schwinger and Weisskopf, and Oppenheimer have suggested that a possible explanation might be the shift of energy levels by the inter-action of the electron with the radiation field. This shift comes out infinite in all existing theories, and *has therefore always been ignored.*[18]

Or consider Lamb's comment on Bethe in his Nobel Prize address:

A month later [after 'the fine structure deviations were definitely established experimentally' by Lamb and Retherford], Bethe found that quantum electrodynamics had really hidden behind its divergences a physical content that was in very close agreement with the micro-wave observations.[19]

Now we attend to the imaginary terms because they have real 'physical content . . . in very close agreement with micro-wave observations'. But until Lamb's experiments they were just mathematical debris which represented nothing of physical significance and were, correctly, omitted.

2.2.   *The Lamb Shift in the Ground State*

The details of the second example are in G. S. Agarwal's monograph on spontaneous emission.[20] Recall that there are two common methods for treating spontaneous emission. The first is the Weisskopf–Wigner method, and the second is via a Markov approximation, leading to a master equation or a Langevin equation, analogous to those used in classical statistical mechanics. As Agarwal stresses, one reason for preferring the newer statistical approach is that it allows us to derive the Lamb shift in the ground state, which is not pre-dicted by the Weisskopf–Wigner method even after that method has been amended to obtain the shift in the excited state. But we can derive the ground state shift only if we are careful about how we use the rotating wave approximation.

---

[18] Hans Bethe, 'The Electromagnetic Shift of Energy Levels', *Physics Review*, 72 (1947), p. 339, italics added.

[19] Willis E. Lamb, Jr., 1955 Nobel Prize Address, *Science* 123 (1956), p. 442.

[20] G. S. Agarwal, *Quantum Statistical Theories of Spontaneous Emission and their Relation to Other Approaches* (Berlin: Springer-Verlag, 1974). See Chapter 10 and Appendix A.

The rotating wave approximation is used when the inter-action between radiation and matter is weak. In weak inter-actions, such as those that give rise to spontaneous emission, the atoms and field can be seen as almost separate systems, so that energy lost by the atoms will be found in the field, and vice versa. Thus virtual transitions in which both the atom and the field simultaneously gain or lose a quantum of energy will have negligible effects. The rotating wave approximation ignores these effects. When the coupling is weak, the terms which represent virtual transitions vary as $\exp\{\pm i(\omega + \omega_k)t\}$ ($\hbar\omega = E_m - E_n$, for energies levels E_m and E_n of the atom; ω_k is a mode frequency of the field). Energy conserving transitions vary as $\exp\{\pm i(\omega - \omega_k)t\}$. For optical frequencies ω_k is large. Thus for ordinary times of observation the $\exp\{\pm i(\omega + \omega_k)t\}$ terms oscillate rapidly, and will average approximately to zero. The approximation is called a 'rotating-wave' approximation because it retains only terms in which the atom and field waves 'rotate together'.

The statistical treatments which give rise to the master equation make an essential use of a Markov approximation. In the last section I outlined one standard way to derive the Pauli equations, which are the master equations relevant for spontaneous emission. But according to Agarwal, there are two ways to carry through such a derivation, and the results are significantly different depending on where we apply the rotating wave approximation. On the one hand, we can begin with the full Hamiltonian for the interaction of the orbiting electron with the electromagnetic field, and drop from this Hamiltonian the 'counter-rotating' terms which represent virtual transitions, to obtain a shortened approximate Hamiltonian which Agarwal numbers (2.24). Then, following through steps like those described above, one obtains a version of the master equation—Agarwal's equation A.7. Alternatively, we can use the full Hamiltonian throughout, dropping the counter-rotating terms only in the last step. This gives us Agarwal's equation (A.6).

What difference do the two methods make to the Lamb shift? Agarwal reports:

The shift of the ground state is missing from (A.7), mainly due to the virtual transitions which are automatically excluded from the

Hamiltonian (2.24). The master equation (A.6) obtained by making RWA (the rotating wave approximation) on the master equation rather than on the Hamiltonian does include the shift of the ground state. These remarks make it clear that RWA on the original Hamiltonian is not the same as RWA on the master equation and that one should make RWA on the final equations of motion.[21]

The rotating wave approximation is justified in cases of spontaneous emission by the weakness of the coupling between the atom and the field. But no further features of the interaction determine whether we should apply the approximation to the original Hamiltonian, or whether instead we should apply it to the master equation. Agarwal applies it to the master equation, and he is thus able to derive a Lamb shift in the ground state. But his derivation does not show that the Schroedinger equation dictates a Lamb shift for a two-level atom in weak interaction with an electro-magnetic field. The shift is consistent with what the equation says about weak interactions, but it does not follow from it.

This kind of situation is even more strikingly illustrated if we try to calculate the values of the shifts for the two-level atoms. Lamb and Retherford's experiments, for example, measured the value of the shift for the $2S$ state in hydrogen to be 1057 mega-cycles per second. We can 'derive' a result very close to this in quantum electrodynamics using the technique of mass renormalization for the electron. But the derivation is notorious: the exact details, in which infinities are subtracted from each other in just the right way to produce a convergent result, are completely *ad hoc*, and yet the quantitative results that they yield are inordinately accurate.

The realist has a defence ready. I say that the Schroedinger equation does not make a claim about whether there is or is not a Lamb shift in the circumstances described. But the realist will reply that I have not described the circumstances as fully as possible. The rotating wave approximations depend on the fact that the interaction between the field and the atom is 'weak'; but if realism is correct, there is a precise answer to the question 'How weak?' The atom-field inter-action will have a precise quantitative representation that

---

[21] Agarwal, op. cit., p. 116.

can in principle be written into the Schroedinger equation. The exact solution to the equation with that term written in will either contain a Lamb shift, or it will not, and *that* is what the Schroedinger equation says about a Lamb shift in cases of spontaneous emission.

This defence allows me to state more exactly the aims of this essay. I do not argue against taking the fundamental laws as true, but try only to counter the most persuasive argument in favour of doing so. As a positive argument in favour of realism, the realist's defence, which plumps for exact solutions and rigorous derivations, gets the logic of the debate out of order. I begin with the challenge, 'Why should we assume that the short abstract equations, which form the core of our fundamental theories, are true at all, even "true enough for now"?' The realist replies 'Because the fundamental equations are so successful at explaining a variety of messy and complicated phenomenological laws'. I say 'Tell me more about explanation. What is there in the explanatory relation which guarantees that the truth of the explanandum argues for the truth of the explanans?' The realist's answer, at least on the plausible generic-specific account, is that the fundamental laws say the same thing as the phenomenological laws which are explained, but the explanatory laws are more abstract and general. Hence the truth of the phenomenological laws is good evidence for the truth of the fundamental laws. I reply to this, 'Why should we think the fundamental and the phenomenological laws say the same thing about the specific situations studied?' and the realist responds, 'Just look at scientific practice. There you will see that phenomenological laws are deducible from the more fundamental laws that explain them, once a description of the circumstances is given.'

We have just been looking in detail at cases of scientific practice, cases which I think are fairly typical. Realists can indeed put a gloss on these examples that brings them into line with realist assumptions: rigorous solutions to exact equations might possibly reproduce the correct phenomenological laws with no ambiguity 'when the right equations are found'. But the reason for believing in this gloss is not the practice itself, which we have been looking at, but rather the realist metaphysics, which I began by challenging. Examples of the sort we have considered here could at best be thought consistent with realist assumptions, but they do not argue for them.

3. CONCLUSION

Fundamental laws are supposed by many to determine what phenomenological laws are true. If the primary argument for this view is the practical explanatory success of the fundamental laws, the conclusion should be just the reverse. We have a very large number of phenomenological laws in all areas of applied physics and engineering that give highly accurate, detailed descriptions of what happens in realistic situations. In an explanatory treatment these are derived from fundamental laws only by a long series of approximations and emendations. Almost always the emendations improve on the dictates of the fundamental law; and even where the fundamental laws are kept in their original form, the steps of the derivation are frequently not dictated by the facts. This makes serious trouble for the D-N model, the generic-specific account, and the view that fundamental laws are better. When it comes to describing the real world, phenomenological laws win out.

Essay 7

Fitting Facts to Equations

0. INTRODUCTION

At H. P. Grice's seminar on metaphysics in the summer of 1975, we discussed Aristotle's categories. I argued then that the category of quantity was empty; there were no quantities in nature—no attributes with exact numerical values of which it could be said that they were either precisely equal or unequal to each other. I was thinking particularly about physics, and the idea I had was like the one I have been defending in these essays, that the real content of our theories in physics is in the detailed causal knowledge they provide of concrete processes in real materials. I thought that these causal relations would hold only between qualities and not between quantities. Nevertheless I recognized that real materials are composed of real atoms and molecules with numerically specific masses, and spins, and charges; that atoms and molecules behave in the way that they do because of their masses, spins, and charges; and that our theoretical analyses of the causal processes they are involved in yield precise numerical calculations of other quantities, such as line shapes in spectroscopy or transport coefficients in statistical mechanics.

Why then did I want to claim that these processes were essentially qualitative? It was because our knowledge about them, while detailed and precise, could not be expressed in simple quantitative equations of the kind that I studied in theoretical physics. The distinction I wanted, it turns out, was not that between the qualitative and the quantitative, but rather the distinction between the tidy and simple mathematical equations of abstract theory, and the intricate and messy descriptions, in either words or formulae, which express our knowledge of what happens in real systems made of real materials, like helium-neon lasers or turbo-jet engines. We may use the fundamental equations of physics to calculate precise quantitative facts about real situations, but as I have urged in earlier essays, abstract fundamental laws are nothing like the complicated, messy laws which describe reality. I no longer want to urge, as I did in the summer seminar, that there are no quantities in nature, but rather that nature is not governed by simple quantitative equations of the kind we write in our fundamental theories.

My basic view is that fundamental equations do not govern objects in reality; they govern only objects in models. The second half of this thesis grew out of another Grice seminar on metaphysics not long after. In the second seminar we talked about pretences, fictions, surrogates, and the like; and Grice asked about various theoretical claims in physics, where should we put the 'as if' operator: helium gas behaves *as if* it is a collection of molecules which interact only on collision? Or, helium gas is composed of molecules which behave *as if* they interact only on collision? Or . . . ?

Again, I wanted to make apparently conflicting claims. There are well-known cases in which the 'as if' operator should certainly go all the way in front: the radiating molecules in an ammonia maser behave *as if* they are classical electron oscillators. (We will see more of this in the last essay.) How closely spaced are the oscillators in the maser cavity? This realistic question is absurd; classical electron oscillators are themselves a mere theoretical construct. What goes on in a real quantum atom is remarkably like the theoretical prescriptions for a classical electron oscillator. The oscillators replicate the behaviour of real atoms; but still, as laser specialist Anthony Siegman remarked in his laser engineering class, 'I wouldn't know where to get myself a bagful of them'.[1]

The classical electron oscillators are undoubted fictions. But, even in cases where the theoretical entities are more robust, I still wanted to put the 'as if' operator all the way in front. For example, a helium-neon laser behaves *as if* it is a collection of three-level atoms in interaction with a single damped mode of a quantized field, coupled to a pumping and damping reservoir. But in doing so, I did not want to deny that the laser cavities contain three-level

[1] Anthony Siegman, 'Lasers' (Electrical Engineering 231), Stanford University, Autumn Term 1981-2.

atoms or that a single mode of the electromagnetic field is dominant. I wanted both to acknowledge these existential facts and yet to locate the operator at the very beginning.

It seems now that I had conflicting views about how to treat this kind of case because I was conflating two functions which the operator could serve. On the one hand, putting things to the left of the operator is a sign of our existential commitment. A helium-neon laser *is* a collection of three-level atoms . . . But putting things on the right serves a different function. Commonly in physics what appears on the right is just what we need to know to begin our mathematical treatment. The description on the right is the kind of description for which the theory provides an equation. We say that a 'real quantum atom' behaves like a classical electron oscillator; already the theory tells us what equation is obeyed by a classical electron oscillator. Similarly, the long description I gave above of a laser as a collection of three-level atoms also tells us a specific equation to write down, in this case an equation called the Fokker–Planck equation; and there are other descriptions of gas lasers which go with other equations. We frequently, for example, treat the laser as a van der Pol oscillator, and then the appropriate equation would be the one which B. van der Pol developed in 1920 for the triode oscillator.

Contrary to my initial assumption I now see that the two functions of the 'as if' operator are quite distinct. Giving a description to which the theory ties an equation can be relatively independent of expressing existential commitment. Both treatments of the laser which I mentioned assume that the helium-neon laser contains a large number of three-level neon atoms mixed with a much greater number of helium atoms, in interaction almost entirely with a single mode of the electromagnetic field. Similarly, when an experimentalist tells us of a single mode of a CW GaAs (gallium arsenide) laser that 'below threshold the mode emits noise *like a* narrow band black body source; above threshold its noise is *characteristic of* a quieted amplitude stabilized oscillator',[2] he is telling us not that the make-up of the laser

---

[2] T. A. Armstrong and A. W. Smith, 'Intensity Fluctuations in a GaAs Laser', *Physical Review Letters* 14 (1965), p. 68, italics added.

has changed but rather that its intensity fluctuations follow from different equations above and below threshold. In these cases what goes on the right of the 'as if' operator does not depend on what we take to be real and what we take to be fictional. Rather it depends on what description we need to know in order to write down the equation that starts our mathematical treatment.

The views I urged in the two seminars go hand-in-hand. It is because the two functions of the 'as if' operator are independent that the fundamental equations of our theories cannot be taken to govern objects in reality. When we use the operator to express existential commitment, we should describe on the left everything we take to be real. From a first, naïve point of view, to serve the second function we should just move everything from the left of the operator to the right. To get a description from which we can write down an equation, we should simply report what we take to be the case.

But that is not how it works. The theory has a very limited stock of principles for getting from descriptions to equations, and the principles require information of a very particular kind, structured in a very particular way. The descriptions that go on the left—the descriptions that tell what there is—are chosen for their descriptive adequacy. But the 'descriptions' on the right—the descriptions that give rise to equations—must be chosen in large part for their mathematical features. This is characteristic of mathematical physics. The descriptions that best describe are generally not the ones to which equations attach. This is the thesis that I will develop in the remaining sections of this paper.

1. TWO STAGES OF THEORY ENTRY

Let us begin by discussing bridge principles. On what Fred Suppe has dubbed the 'conventional view of theories',[3] championed by Hempel, Grünbaum, Nagel, and others in the tradition of logical empiricism, the propositions of a theory are of two kinds: internal principles and bridge

[3] In Frederick Suppe, *The Structure of Scientific Theories* (Urbana: University of Illinois Press, 1977).

principles. The internal principles present the content of the theory, the laws that tell how the entities and processes of the theory behave. The bridge principles are supposed to tie the theory to aspects of reality more readily accessible to us. At first the bridge principles were thought to link the descriptions of the theory with some kind of observation reports. But with the breakdown of the theory–observation distinction, the bridge principles were required only to link the theory with a vocabulary that was 'antecedently understood'.

The network of internal principles and bridge principles is supposed to secure the deductive character of scientific explanation. To explain why lasers amplify light signals, one starts with a description in the antecedent vocabulary of how a laser is constructed. A bridge principle matches this with a description couched in the language of the quantum theory. The internal principles of quantum mechanics predict what should happen in situations meeting this theoretical description, and a second bridge principle carries the results back into a proposition describing the observed amplification. The explanation is deductive because each of the steps is justified by a principle deemed necessary by the theory, either a bridge principle or an internal principle.

Recently, however, Hempel has begun to doubt that explanations of this kind are truly deductive.[4] The fault lies with the bridge principles, which are in general far from exceptionless, and hence lack the requisite necessity. A heavy bar attracts iron filings. Is it thus magnetic? Not necessarily: we can never be sure that we have succeeded in ruling out all other explanations. A magnet will, with surety, attract iron filings only if all the attendant circumstances are right. Bridge principles, Hempel concludes, do not have the character of universal laws; they hold only for the most part, or when circumstances are sufficiently ideal.

I think the situation is both much better and much worse than Hempel pictures. If the right kinds of descriptions are given to the phenomena under study, the theory will

---

[4] C. G. Hempel, in a paper at the conference 'The Limits of Deductivity', University of Pittsburgh, Autumn 1979.

tell us what mathematical description to use and the principles that make this link are as necessary and exceptionless in the theory as the internal principles themselves. But the 'right kind of description' for assigning an equation is seldom, if ever, a 'true description' of the phenomenon studied; and there are few formal principles for getting from 'true descriptions' to the kind of description that entails an equation. There are just rules of thumb, good sense, and, ultimately, the requirement that the equation we end up with must do the job.

Theory entry proceeds in two stages. I imagine that we begin by writing down everything we know about the system under study, a gross exaggeration, but one which will help to make the point. This is the *unprepared description*—it is the description that goes to the left of the 'as if' operator when the operator is used to express existential commitment. The unprepared description contains any information we think relevant, in whatever form we have available. There is no theory–observation distinction here. We write down whatever information we have: we may know that the electrons in the beam are all spin up because we have been at pains to prepare them that way; or we may write down the engineering specifications for the construction of the end mirrors of a helium-neon laser; and we may also know that the cavity is filled with three-level helium atoms. The unprepared description may well use the language and the concepts of the theory, but it is not constrained by any of the mathematical needs of the theory.

At the first stage of theory entry we prepare the description: we present the phenomenon in a way that will bring it into the theory. The most apparent need is to write down a description to which the theory matches an equation. But to solve the equations we will have to know what boundary conditions can be used, what approximation procedures are valid, and the like. So the prepared descriptions must give information that specifies these as well. For example, we may describe the walls of the laser cavity and their surroundings as a *reservoir* (a system with a large number of resonant modes). This means that the laser has no memory. Formally, when we get to the derivation, we can make a Markov approximation. (Recall the discussion in Essay 6.)

This first stage of theory entry is informal. There may be better and worse attempts, and a good deal of practical wisdom helps, but no principles of the theory tell us how we are to prepare the description. We do not look to a bridge principle to tell us what is the right way to take the facts from our antecedent, unprepared description, and to express them in a way that will meet the mathematical needs of the theory. The check on correctness at this stage is not how well we have represented in the theory the facts we know outside the theory, but only how successful the ultimate mathematical treatment will be.

This is in sharp contrast with the second stage of theory entry, where principles of the theory look at the prepared description and dictate equations, boundary conditions, and approximations. Shall we treat a CW GaAs laser below threshold as a 'narrow band black body source' rather than the 'quieted stabilized oscillator' that models it above threshold? Quantum theory does not answer. But once we have decided to describe it as a narrow band black body source, the principles of the theory tell what equations will govern it. So we do have bridge principles, and the bridge principles are no more nor less universal than any of the other principles. But they govern only the second stage of theory entry. At the first stage there are no theoretical principles at all—only rules of thumb and the prospect of a good prediction.

This of course is a highly idealized description. Theories are always improving and expanding, and an interesting new treatment may offer a totally new bridge principle. But Hempel's original account was equally idealized; it always looked at the theory as it stood after the explanation had been adopted. I propose to think about it in the same way. In the next section I want to illustrate some bridge principles, and I shall describe the two stages of theory entry with some examples from quantum mechanics.

2. SOME MODEL BRIDGE PRINCIPLES

If we look at typical formalizations of quantum mechanics,[5] it seems that the fundamental principles do divide into internal principles and bridge principles, as the conventional view of theories maintains. The central internal principle is Schroedinger's equation. The Schroedinger equation tells how systems, subject to various forces, evolve in time. Actually the forces are not literally mentioned in the equation since quantum mechanics is based on William Hamilton's formulation of classical mechanics, which focuses not on forces, but on energies. In the standard presentation, the Schroedinger equation tells how a quantum system evolves in time when the *Hamiltonian* of the system is known, where the Hamiltonian is a mathematical representation of the kinetic and potential energies for the system. Conservation principles, like the conservation of energy, momentum, or parity, may also appear as internal principles in such a formalization. (On the other hand, they may not, despite the fact that they are of fundamental importance, because these principles can often be derived from other basic principles.)

The second class of principles provide schemata for getting into and out of the mathematical language of the theory: states are to be represented by vectors; observable quantities are represented by operators; and the average value of a given quantity in a given state is represented by a certain product involving the appropriate operator and vector. So far, all looks good for the conventional view of theories.

But notice: one may know all of this and not know any quantum mechanics. In a good undergraduate text these two sets of principles are covered in one short chapter. It is true that the Schroedinger equation tells how a quantum system evolves subject to the Hamiltonian; but to do quantum mechanics, one has to know how to pick the Hamiltonian.

---

[5] See, for example, the formalization in Chapter 5, 'Development of the Formalism of Wave Mechanics and its Interpretation', in Albert Messiah's highly respected text, *Quantum Mechanics* (Amsterdam: North-Holland, 1969) or the formal axiomatization in George W. Mackey's *Mathematical Foundations of Quantum Mechanics* (New York: W. A. Benjamin, 1963). '

The principles that tell us how to do so are the real bridge principles of quantum mechanics. These give content to the theory, and these are what beginning students spend the bulk of their time learning.

If the coventional view were right, students should be at work learning bridge principles with mathematical formulae on one end and descriptions of real things on the other. Good textbooks for advanced undergraduates would be full of discussions of concrete situations and of the Hamiltonians which describe them. There might be simplifications and idealization for pedagogical purposes; nevertheless, there should be mention of concrete things made of the materials of the real world. This is strikingly absent. Generally there is no word of any material substance. Instead one learns the bridge principles of quantum mechanics by learning a sequence of model Hamiltonians. I call them 'model Hamiltonians' because they fit only highly fictionalized objects. Here is a list of examples. I culled it from two texts, both called *Quantum Mechanics*, one by Albert Messiah[6] and the other by Eugen Merzbacher.[7] This list covers what one would study in just about any good senior level course on quantum mechanics. We learn Hamiltonians for:

free particle motion, including
    the free particle in one dimension,
    the free particle in three dimensions,
    the particle in a box;
the linear harmonic oscillator;
piecewise constant potentials, including
    the square well,
    the potential step,
    the periodic potential,
    the Coulomb potential;
'the hydrogen atom';
diatomic molecules;
central potential scattering;

[6] Messiah, op. cit.
[7] Eugen Merzbacher, *Quantum Mecahnics*, Second Edition (New York: John Wiley & Sons, 1970).

and eventually, the foundation of all laser theory,

the electron in interaction with the electromagnetic field.

There is one real material mentioned in this list—hydrogen. In fact this case provides a striking illustration of my point, and not a counterexample against it. The Hamiltonian we learn here is not that for any real hydrogen atom. Real hydrogen atoms appear in an environment, in a very cold tank for example, or on a benzene molecule; and the effects of the environment must be reflected in the Hamiltonian. What we study instead is a hypothetically isolated atom. We hope that later we will be able to piece together this Hamiltonian with others to duplicate the circumstances of an atom in its real situation.

But this is not the most striking omission. In his section titled 'The Hydrogen Atom', Messiah proposes a particular Hamiltonian and uses it to provide a solution for the energy spectrum of hydrogen. He says:

This spectrum is just the one predicted by the Old Quantum Theory; its excellent agreement with the experimental spectrum was already pointed out. To be more precise, the theory correctly accounts for the position of the spectral lines but not for their fine structure. Its essential shortcoming is to be a non-relativistic theory . . . [Also] the Schroedinger theory does not take the electron spin into account.[8]

These are critical omissions. The discovery and the account of the fine structure of hydrogen were significant events in quantum mechanics for the reasons Messiah mentions. Fine structure teaches important lessons both about relativity and about the intrinsic spin of the electron.

The passage quoted above appears about three quarters of the way through Volume I. About the same distance into Volume II, Messiah again has a section called 'The Hydrogen Atom'. There he uses the relativistic theory of the Dirac electron. Even the second treatment is not true to the real hydrogen atom. The reasons are familiar from our discussion of the Lamb shift in Essay 6. Here is what Messiah himself says:

---

[8] Messiah, op. cit., p. 419.

The experimental results on the fine structure of the hydrogen atom and hydrogen-like atoms (notably He+) are in broad agreement with these predictions.

However, the agreement is not perfect. The largest discrepancy is observed in the fine structure of the $n = 2$ levels of the hydrogen atom. In the non-relativistic approximation, the three levels $2s_{1/2}$, $2p_{1/2}$, and $2p_{3/2}$ are equal. In the Dirac theory, the levels $2s_{1/2}$, and $2p_{1/2}$ are still equal, while the $2p_{3/2}$ level is slightly lower (the separation is of the order of 10^{-4}eV). The level distance from $2p_{3/2}$ to $2p_{1/2}$ agrees with the theory but the level $2s_{1/2}$ is lower than the level $2p_{1/2}$, and the distance from $2s_{1/2}$ to $2p_{1/2}$ is equal to about a tenth of the distance from $2p_{3/2}$ to $2p_{1/2}$. This effect is known as the Lamb shift. To explain it, we need a rigorous treatment of the interaction between the electron, the proton and the quantized electromagnetic field; in the Dirac theory one retains only the Coulomb potential which is the main term in that interaction; the Lamb shift represents 'radiative corrections' to this approximation.[9]

We know from our earlier discussion that the treatment of these 'radiative corrections' for the hydrogen spectrum is no simple matter.

The last sentence of Messiah's remark is telling. The two sections are both titled 'The Hydrogen Atom' but in neither are we given a Hamiltonian for real hydrogen atoms, even if we abstract from the environment. Instead, we are taught how to write the *Coulomb potential between an electron and a proton*, in the first case non-relativistically and in the second, relativistically. Messiah says so himself: 'The simplest system of two bodies with a Coulomb interaction is the hydrogen atom'.[10] 'The hydrogen atom' on our list is just a name for a two-body system where only the Coulomb force is relevant. Even if the system stood alone in the universe, we could not strip away the spin from the electron. Even less could we eliminate the electromagnetic field, for it gives rise to the Lamb shift even when no photons are present. This two-body system, which we call 'the hydrogen atom', is a mere mental construct.

Messiah's is of course an elementary text, intended for seniors or for beginning graduate students. Perhaps we are looking at versions of the theory that are too elementary?

---

[9] Ibid., pp. 932–3.   [10] Ibid., p. 412.

Do not more sophisticated treatments—journal articles, research reports, and the like—provide a wealth of different, more involved bridge principles that link the theory to more realistic descriptions? I am going to argue in the next chapter that the answer to this question is *no*. There are some more complicated bridge principles; and of course the theory is always growing, adding both to its internal principles and its bridge principles. But at heart the theory works by piecing together in original ways a small number of familiar principles, adding corrections where necessary. This is how it should work. The aim is to cover a wide variety of different phenomena with a small number of principles, and that includes the bridge principles as well as the internal principles. It is no theory that needs a new Hamiltonian for each new physical circumstance. The explanatory power of quantum theory comes from its ability to deploy a small number of well-understood Hamiltonians to cover a wide range of cases. But this explanatory power has its price. If we limit the number of Hamiltonians, that is going to constrain our abilities to represent situations realistically. This is why our prepared descriptions lie.

I will take up these remarks about bridge principles again in the next chapter. Here I want to proceed in a different way. I claim that in general we will have to distort the true picture of what happens if we want to fit it into the highly constrained structures of our mathematical theories. I think there is a nice analogy that can help us see why this is so. That is the topic of the next section.

3. PHYSICS AS THEATRE

I will present first an analogy and then an example. We begin with Thucydides' views on how to write history:

XXII. As to the speeches that were made by different men, either when they were about to begin the war or when they were already engaged therein, it has been difficult to recall with strict accuracy the words actually spoken, both for me as regards that which I myself heard, and for those who from various other sources have brought me reports. Therefore the speeches are given in the language in which, as it seemed to me, the several speakers would express, on the subjects under consideration, the sentiments most befitting the occasion, though at the same time I have adhered as closely as possible to the general sense of what was actually said.[11]

Imagine that we want to stage a given historical episode. We are primarily interested in teaching a moral about the motives and behaviour of the participants. But we would also like the drama to be as realistic as possible. In general we will not be able simply to 'rerun' the episode over again, but this time on the stage. The original episode would have to have a remarkable unity of time and space to make that possible. There are plenty of other constraints as well. These will force us to make first one distortion, then another to compensate. Here is a trivial example. Imagine that two of the participants had a secret conversation in the corner of the room. If the actors whisper together, the audience will not be able to hear them. So the other characters must be moved off the stage, and then back on again. But in reality everyone stayed in the same place throughout. In these cases we are in the position of Thucydides. We cannot replicate what the characters actually said and did. Nor is it essential that we do so. We need only adhere 'as closely as possible to the general sense of what was actually said'.

Physics is like that. It is important that the models we construct allow us to draw the right conclusions about the behaviour of the phenomena and their causes. But it is not essential that the models accurately describe everything that actually happens; and in general it will not be possible for them to do so, and for much the same reasons. The requirements of the theory constrain what can be literally represented. This does not mean that the right lessons cannot be drawn. Adjustments are made where literal correctness does not matter very much in order to get the correct effects where we want them; and very often, as in the staging example, one distortion is put right by another. That is why it often seems misleading to say that a particular aspect of a model is false to reality: given the other constraints that is just the way to restore the representation.

Here is a very simple example of how the operation of

[11] Thucydides, *The Peloponnesian War*, Vol. 1, trans. Charles Forster Smith (New York: G. P. Putnam's Sons, 1923), p. 39.

constraints can cause us to set down a false description in physics. In quantum mechanics free particles are represented by plane waves—functions that look like sines or cosines, stretching to infinity in both directions. This is the representation that is dictated by the Schroedinger equation, given the conventional Hamiltonian for a free particle. So far there need be nothing wrong with a wave like that. But quantum mechanics has another constraint as well: the square of the wave at a point is supposed to represent the probability that the particle is located at that point. So the integral of the square over all space must equal one. But that is impossible if the wave, like a sine or cosine, goes all the way to infinity.

There are two common solutions to this problem. One is to use a Dirac delta function. These functions are a great help to physics, and generalized function theory now explains how they work. But they side-step rather than solve the problem. Using the delta function is really to give up the requirement that the probabilities themselves integrate to one. Merzbacher, for instance, says 'Since normalization of $\int \Psi^* \Psi$ to unity is out the question for infinite plane waves, we must decide on an *alternative* normalization for these functions. A convenient tool in the discussion of such wave functions is the delta function'.[12] I have thus always preferred the second solution.

This solution is called 'box normalization'. In the model we assume that the particle is in a very, very large box, and that the wave disappears entirely at the edges of this box. To get the wave to go to zero we must assume that the potential there—very, very far away from anything we are interested in—is infinite. Here is what Merzbacher says in defence of this assumption:

The eigenfunctions are not quadratically integrable over all space. It is therefore impossible to speak of absolute probabilities and of expectation values for physical quantities in such a state. One way of avoiding this predicament would be to recognize the fact that physically no particle is ever absolutely free and that there is inevitably some confinement, the enclosure being for instance the wall

---

[12] Merzbacher, op. cit., p. 82, italics added.

of an accelerator tube or of the laboratory. V [the potential] rises to infinity at the boundaries of the enclosure and does then not have the same value everywhere, the eigenfunctions are no longer infinite plane waves, and the eigenvalue spectrum is discrete rather than continuous.[13]

Here is a clear distortion of the truth. The walls may interact with the particle and have some effect on it, but they certainly do not produce an infinite potential.

I think Merzbacher intends us to think of the situation this way. The walls and environment do contain the particle; and in fact the probability is one that the particle will be found in some finite region. The way to get this effect in the model is to set the potential at the walls to infinity. Of course this is not a true description of the potentials that are actually produced by the walls and the environment. But it is not exactly false either. It is just the way to achieve the results in the model that the walls and environment are supposed to achieve in reality. The infinite potential is a good piece of staging.

[13] Ibid., p. 82.

The Simulacrum Account of Explanation

0. INTRODUCTION

We saw in the last chapter that the bridge principles in a theory like quantum mechanics are few in number, and they deal primarily with highly fictionalized descriptions. Why should this be so? Some work of T. S. Kuhn suggests an answer. In his paper, 'A Function for Measurement in the Physical Sciences', and in other papers with the word 'function' in the title, Kuhn tries something like functional explanations of scientific practice. Anthropologists find a people with a peculiar custom. The people themselves give several or maybe no reasons for their custom. But the anthropologist conjectures that the custom remains among these people not only for their avowed reasons, but also because other customs or ecological conditions make it very difficult for their society to survive without it. This then is the 'function' of the custom in question, even though it is not practised with any conscious awareness of that function. Naturally all functional explanations have a dubious logic, but they do often bring out instructive aspects of the custom in question.

Now let us ask what function might be served by having relatively few bridge principles to hand when we are engaged in constructing models of phenomena. Kuhn concludes his paper on measurement by saying he believes 'that the nineteenth century mathematization of physical science produced vastly refined professional criteria for problem selection and that it simultaneously very much increased the effectiveness of professional verification procedures'.[1] I think that something similar is to be said about having a rather small number of bridge principles. The phenomena to be described are endlessly complex. In order to pursue any collective research, a group must be able to delimit the kinds of models that

---

[1] T. S. Kuhn, 'A Function for Measurement in the Physical Sciences', in T. S. Kuhn, *The Essential Tension* (Chicago: University of Chicago Press, 1977), p. 220.

are even contenders. If there were endlessly many possible ways for a particular research community to hook up phenomena with intellectual constructions, model building would be entirely chaotic, and there would be no consensus of shared problems on which to work.

The limitation on bridge principles provides a consensus within which to formulate theoretical explanations and allows for relatively few free parameters in the construction of models. This in turn provides sharp criteria for problem selection. Naturally there may be substantial change in the structure of bridge principles if nothing works; but we hang on to them while we can. It is precisely the existence of relatively few bridge principles that makes possible the construction, evaluation, and elimination of models. This fact appears also to have highly anti-realist side effects. As we have seen, it strongly increases the likelihood that there will be literally incompatible models that all fit the facts so far as the bridge principles can discriminate.

This is just a sketch of a Kuhnian account, but one which I believe is worth pursuing. Borrowing a term from the historians of science, it might be called an 'external' account of why bridge principles need to be limited in number. But it fits nicely with a parallel 'internal' account, one that holds that the limitation on bridge principles is crucial to the explanatory power of the theory. I will argue for this internal account in section 1 of this essay. In section 2 I propose a model of explanation that allows for the paucity of bridge principles and makes plain the role of fictionalized descriptions.

2.   BRIDGE PRINCIPLES AND 'REALISTIC' MODELS

A good theory aims to cover a wide variety of phenomena with as few principles as possible. That includes bridge principles. It is a poor theory that requires a new Hamiltonian for each new physical circumstance. The great explanatory power of quantum mechanics comes from its ability to deploy a small number of well-understood Hamiltonians to cover a broad range of cases, and not from its ability to match each situation one-to-one with a new mathematical representation. That way of proceeding would be crazy.

This is an obvious fact about what theories must be like if they are to be manageable at all. But it has the anti-realist consequences that we have seen. Why have realists not been more troubled by this fact? The answer, I think, is that many realists suppose that nature conspires to limit the number of bridge principles. Only a few bridge principles are needed because only a small number of basic interactions exist in nature. An ideal theory will represent each of the basic interactions; new cases will not require new bridge principles because the representation for complex situations can be constructed from the representations for the basic components.

I think that this is a radically mistaken point of view. First, it is a model of a physics we do not have. That is a familiar point by now. Much worse, it is a model of a physics we do not want. The piecing-together procedure would be unbearably complex. It goes in exactly the wrong direction. The beauty and strength of contemporary physics lies in its ability to give simple treatments with simple models, where at least the behaviour in the model can be understood and the equations can not only be written down but can even be solved in approximation. The harmonic oscillator model is a case in point. It is used repeatedly in quantum mechanics, even when it is difficult to figure out exactly what is supposed to be oscillating: the hydrogen atom is pictured as an oscillating electron; the electromagnetic field as a collection of quantized oscillators; the laser as a van der Pol oscillator; and so forth. The same description deployed again and again gives explanatory power.

It is best to illustrate with a concrete case. In the last essay we looked to elementary texts for examples of bridge principles. Here I will present a more sophisticated example: a quantum theoretical account of a laser. Recall from Essay 4 that there are a variety of methods for treating lasers in quantum mechanics. One is the quantum statistical approach in which a master equation (or a Langevin equation) is derived for the system. This kind of approach is familiar from our discussion in Essay 6 of the Markov approximation for radiating atoms, so this is a good example to choose.

There is a thorough development of this method in William Louisell's *Quantum Statistical Properties of Radiation*.[2] The treatment there is most suitable for a gas laser, such as the early helium-neon laser. Louisell proposes what he calls a 'block diagram'. (See Figure 8.1.) He imagines that the laser consists of three-level atoms in interaction with a quantized electromagnetic field. Before the quantum statistical approach, treatments of lasers were generally semi-classical: the atoms were quantized but the field was not. Louisell also explicitly includes the interaction of both the atoms and the field with a damping reservoir. These two features are important for they allow the derivation of correlations among the emitted photons which are difficult to duplicate in the earlier semi-classical approaches.

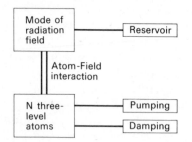

FIG. 8.1. Block diagram of laser model. (*Source*: Louisell, *Quantum Statistical Properties of Radiation*.)

In Essay 6 I talked briefly about idealizations that are hard to eliminate at the theoretical level. Here is a good illustration. Louisell supposes that the atoms are uniformly distributed, N per unit volume, and that they do not interact with each other: 'They are coupled to each other only through their atom-field interaction'.[3] In reality the atoms do interact, though this does not have much effect on the performance of the laser. The missing effects can sometimes

[2] W. H. Louisell, *Quantum Statistical Properties of Radiation* (New York: John Wiley & Sons, 1973), Ch. 9.
[3] Ibid., p. 469.

be corrected for, but this is done piece-meal when the theory is applied and not by adding terms to the fundamental Hamiltonian given in the theoretical treatment.

Louisell's equation for the system represented by his block diagram consists of three parts. I will write it here just for reference:

$$\frac{\partial S}{\partial t} = \frac{1}{i\hbar} [S, H_0 + W] + \left(\frac{\partial S}{\partial t}\right)_F + \left(\frac{\partial S}{\partial t}\right)_A.$$

He tells us of this equation that 'the first terms describe the causal behavior. The second term describes the interaction of the field mode with its reservoir . . . The last term describes the interaction of the atoms with their pumping and damping reservoirs'.[4] This equation is deceptively simple because it is still just a schema. We do not yet know how W, $(\partial S/\partial t)_F$, and so forth are to be represented for the block laser. This is where bridge principles enter. A page and a half later when these variables have been filled in, this simple-seeming equation will take twelve lines of text for Louisell to write.

The first term is supposed to represent the 'causal be-haviour', in contrast with the last two terms. Another com-mon way of expressing this contrast would be to say: the atom-field interaction is represented realistically, but the terms for the reservoir interactions are just phenomenological. Louisell's method of expression is better because it is nar-rower. Physicists use 'realistic' in a variety of senses. One common sense contrasts 'realistic' with 'idealized'. This sense concerns the relation between the model and the situation depicted in the model: how well do the prepared and the unprepared descriptions match? We have seen that Louisell's treatment of the atoms is highly idealized. So too are other aspects of his block diagram. In *this* sense, Louisell's model for the 'causal behaviour' is not very realistic.

There is another, different way in which physicists use the word 'realistic'. I will illustrate with three examples. The first example comes from the laser engineering course I referred to in the last eassy.[5] After several lectures on

[4] Ibid., p. 470.
[5] Anthony Siegman, 'Lasers' (Electrical Engineering 231), Stanford Uni-versity, Autumn term 1981–2.

classical electron oscillators, Professor Anthony Siegman announced that he was ready to talk about the lasing medium in a real laser. I thought he was going to teach about ruby rods—that ruby is chromium doped sapphire, that the 3+ chromium ions are interspersed randomly at low densities throughout the sapphire lattice, and that an ordinary electric discharge is used to excite the chromium ions and bring about a population inversion. Instead he began, 'Consider a collection of two-level atoms.' In a sense he started to talk about real lasers—a laser medium is indeed composed of quantized systems like atoms, and not of the fictitious electron oscillators. But in another sense he did not: two-level atoms are but a crude stand-in for the intricate and variegated structure of real laser materials.

The second example comes from a conversation with my colleague Francis Everitt, an experimental physicist whom I have mentioned before in connection with his historical work on James Clerk Maxwell. In the last essay we saw that a laser can be treated by van der Pol's equation: in a variety of ways the laser will behave like a triode oscillator in a d.c. circuit. In talking with Everitt I contrasted this description with Louisell's. Louisell mentions real components of the laser, like the atoms and the field. I took Louisell's to be the more realistic description. Everitt agreed. But he added, 'The reservoir is still only a model.' In Louisell's diagram the damping reservoir represents the walls of the cavity and the room in which it is housed. The three-level atoms represent the lasing medium. In what sense is the reservoir, unlike the atoms, just a model?

The third example gives an explicit clue. In the text *Quantum Optics* John Klauder and E. C. G. Sudarshan report, 'A number of authors have treated idealized interacting systems as models for lasers'.[6] Louisell is an example. Although highly idealized, the Louisell model is still realistic in a way in which the models of Klauder and Sudarshan are not. They themselves describe their models as 'phenomenological'. What do they mean? They say that their models are phenomenological because the models 'work

[6] J. Klauder and E. C. G. Sudarshan, *Quantum Optics* (New York: Benjamin, 1968), p. 234.

directly on the state . . . as a function of time and do not derive it as a solution to a Hamiltonian'.[7] Recall that the Hamiltonian goes into the Schroedinger equation and determines the time evolution of the state. It represents the energies which guide the behaviour of the system. Sudarshan and Klauder aim to get the right state; but they write down this state directly, *post hoc*, with an eye to the behaviour it is supposed to predict. They do not write down a Schroedinger equation and derive the state as a solution to it; and thus they do not show what energies produce the state. Their treatment is unrealistic *from the point of view of the explanatory theory*. It gives a theoretical description of the behaviour, but nothing in the model shows what gives rise to this behaviour.

Look back now to the damping reservoir, and recall our discussion of atomic radiation in Essay 6. The effect of a damping reservoir is to bring about an irreversible change in the system which couples to it. Information which goes into the reservoir gets lost there, and the memory of the system is erased. The reservoir is a way of representing the walls of the cavity and of the surrounding environment. But it is like the proverbial black box. It generates the effects that the walls are supposed to have, but there is no representation of the method by which the walls produce these effects. No description is given of how the walls are made up, or of what gives rise to the formal characteristics that reservoirs must have to bring about damping. This constrasts with Siegman's treatment of the lasing medium. Two-level atoms are not very much like chromium ions in a ruby laser. But they do give rise to equations in accord with established explanatory principles and not in an *ad hoc* way.

The two senses of 'realistic' act at different levels. The first bears on the relation between the model and the world. The model is realistic if it presents an accurate picture of the situation modelled: it describes the real constituents of the system—the substances and fields that make it up— and ascribes to them characteristics and relations that actually obtain. The second sense bears on the relation between the

[7] Ibid., p. 226.

model and the mathematics. A fundamental theory must supply a criterion for what is to count as explanatory. Relative to that criterion the model is realistic if it explains the mathematical representation.

The two senses of realistic are nicely illustrated in Louisell's treatment. We have already seen that Louisell's model is only quasi-realistic in the first sense. It describes the significant components, but the features it ascribes to them are a caricature of those in reality. The model is both realistic and unrealistic in the second sense as well. The first term in Louisell's equation represents the potential arising from the atom-field interaction which he sets down in the model. That is what he means by saying that it represents the 'causal behaviour'. The reservoir terms are different. They give rise to the right solutions but no concrete mechanisms are supplied in the model to explain them.

The two ways in which a model may be unrealistic are related. Louisell's modelling of the reservoir is unrealistic in the first sense as well as in the second, in part because he does not intend to use the detailed structure of the reservoir to generate his equation. But that is not all there is to it. We say in Essay 6 that if the reservoir is really to do its job in getting the atoms to decay, the time correlations there must be exactly zero. This is an assumption that Louisell makes; but it is highly unrealistic in the first sense. This case is just like the infinite potentials in the last section of the last essay. The conventional Schroedinger theory cannot be fitted exactly to the situation, so we deal with the problem by distorting the situation. But we put the distortion as far away from the system of immediate concern as possible. If we are interested in the atoms only, we can put the distortion in the description of the field, assigning it an infinite number of degrees of freedom. But if we want to study the field as well, the infinite degrees of freedom or the zero time correlations are put into the walls of the cavity and the surrounding environment. And so on.

We learn an important lesson about bridge principles from these considerations. A treatment that is more realistic in the second sense will employ more bridge principles. The quantum statistical approach is a highly sophisticated method for predicting fine details about photon statistics in laser light. Even in an advanced treatment like this, a large part of the work is done by phenomenological terms which minimize the number of bridge principles needed. For example, the phenomenological terms that Louisell employs are from his general theory of damped systems and can be employed again and again independent of how the damping is brought about.

The first term of Louisell's equation also illustrates this point about bridge principles. In the last essay I raised the worry that the bridge principles I discussed there were too elementary to be instructive. Louisell's equation shows that this is not so. Only the first term is a genuine Hamiltonian term matched by a conventional bridge principle with a description of the potential. What Hamiltonian is it? It is just the Hamiltonian for the interaction of an atom with a radiation field, which appeared on our list in Essay 7 and which was developed in a classic paper by Enrico Fermi in 1932. The only change that Louisell makes is to sum the Hamiltonian over all the atoms in the cavity. This bears out my general claim about bridge principles. The success of the quantum statistical treatment does not depend on using novel principles that are highly involved, but rather in using some well-known and well-understood principles in a novel way.

2.   THE SIMULACRUM ACCOUNT OF EXPLANATION

The conventional D-N account supposes that we have explained a phenomenon when we have shown how it follows from a more fundamental law. This requires that the treatments we give for phenomena in physics must certainly be realistic in the first sense, and preferably in the second as well, if they are to serve as explanations. I propose an alternative to the D-N model that brings the philosophic account closer to explanatory practices in physics as I have pictured them. It is based on Duhem's view of explanation, which I sketched in Essay 5, and follows immediately from the discussion of the last section.

The primary aim of this book is to argue against the facticity of fundamental laws. As we saw in the very first essay, one of the chief arguments that realists use in favour of their facticity is their broad explanatory and predictive success. I have been arguing here that the vast majority of successful treatments in physics are not realistic. They are not realistic in the first sense of picturing the phenomena in an accurate way; and even in the second sense, too much realism may be a stop to explanatory power, since the use of 'phenomenological' terms rather than a more detailed 'causal' construction may allow us more readily to deploy known solutions with understood characteristics and thereby to extend the scope of our theory.

If what I say is correct, it calls for a new account of explanation. Recall the discussion of Essay 6. To explain a phenomenon is to find a model that fits it into the basic framework of the theory and that thus allows us to derive analogues for the messy and complicated phenomenological laws which are true of it. The models serve a variety of purposes, and individual models are to be judged according to how well they serve the purpose at hand. In each case we aim to 'see' the phenomenon through the mathematical framework of the theory, but for different problems there are different emphases. We may wish to calculate a particular quantity with great accuracy, or to establish its precise functional relationship to another. We may wish instead to replicate a broader range of behaviour, but with less accuracy. One important thing we sometimes want to do is to lay out the causal processes which bring the phenomena about, and for this purpose it is best to use a model that treats the causally relevant factors as realistically as possible, in both senses of 'realistic'. But this may well preclude treating other factors realistically. We should not be misled into thinking that the most realistic model will serve all purposes best.

In order to stress this 'anti-realistic' aspect of models, I call my view of explanation a 'simulacrum' account. The second definition of 'simulacrum' in the *Oxford English Dictionary* says that a simulacrum is 'something having merely the form or appearance of a certain thing, without

without possessing its substance or proper qualities'.[8] This is just what I have been urging that models in phsyics are like. Is a helium-neon laser really a van der Pol oscillator? Well, it is really a mix of helium and neon atoms, in about the ratio nine to one, enclosed in a cavity with smooth walls and reflecting mirrors at both ends, and hooked up to a device to pump the neon atoms into their excited state. It is not literally a triode oscillator in a d.c. circuit. If we treat it with van der Pol's equation for a triode oscillator, we will be able to replicate a good deal of its behaviour above threshold, and that is our aim. The success of the model depends on how much and how precisely it can replicate what goes on.

A model is a work of fiction. Some properties ascribed to objects in the model will be genuine properties of the objects modelled, but others will be merely properties of convenience. The term 'properties of convenience' was suggested by H. P. Grice, and it is apt. Some of the properties and relations in a model will be real properties, in the sense that other objects in other situations might genuinely have them. But they are introduced into this model as a convenience, to bring the objects modelled into the range of the mathematical theory.

Not all properties of convenience will be real ones. There are the obvious idealizations of physics—infinite potentials, zero time correlations, perfectly rigid rods, and frictionless planes. But it would be a mistake to think entirely in terms of idealizations—of properties which we conceive as limiting cases, to which we can approach closer and closer in reality. For some properties are not even approached in reality. They are pure fictions.

I would want to argue that the probability distributions of classical statistical mechanics are an example. This is a very serious claim, and I only sketch my view here as an illustration. The distributions are essential to the theory— they are what the equations of the theory govern—and the theory itself is extremely powerful, for example in the detailed treatment of fluid flow. Moreover, in some simple special cases the idea of the probability distribution can be

[8] *The Oxford English Dictionary* (Oxford: Oxford University Press, 1933).

operationalized; and the tests support the distributions ascribed by the theory.[9]

Nevertheless, I do not think these distributions are real. Statistical mechanics works in a massive number of highly differentiated and highly complex situations. In the vast majority of these it is incredible to think that there is a true probability distribution for that situation; and proofs that, for certain purposes, one distribution is as good as another, do not go any way to making it plausible that there is one at all. It is better, I think, to see these distributions as fictions, fictions that have a powerful organizing role in any case and that will not mislead us too much even should we take them to be real in the simple cases.

We can illustrate with Maxwell's treatment of the radio-meter, described in the introduction to this book. Maxwell begins with Boltzmann's equation (equation 1, Introduction), which governs the evolution of the velocity distribution of the gas molecules. (This distribution gives the probability, for every possible combination of values for $v, w, x, \ldots$, that the first molecule has velocity v; the second, velocity w; the third velocity x; etc.) Maxwell writes down one of the many functions which solve Boltzmann's equation and he claims that this function is the distribution for 'a medium in which there are inequalities of temperature and velocity' and in which the viscosity varies 'as the first power of the absolute temperature'.[10]

I claim that the medium which Maxwell describes is only a model. It is not the medium which exists in any of the radiometers we find in the toy department of Woolworth's. The radiometers on the shelves in Woolworth's do not have delicate well-tuned features. They cost $2.29. They have a host of causally relevant characteristics besides the two critical ones Maxwell mentions, and they differ in these characteristics from one to another. Some have sizeable convection currents; in others the currents are negligible; probably the co-efficients of friction between vanes and

[9] See, for example, T. K. Roberts and A. R. Miller, *Heat and Thermodynamics* (London: Blackie & Son, 1960).

[10] James Maxwell, 'On Stresses in Rarified Gases arising from Inequalities of Temperature', *The Scientific Papers of James Clark Maxwell*, ed. W. D. Niven (New York: Dover Publications, 1965), pp. 691, 692.

gases differ; as do the conduction rates, the densities of the enclosed gases, and the make-up of the gas itself.

We may be inclined to think that this does not matter much. Maxwell has made a harmless idealization: the effects of the other factors are small, and the true distribution in each Woolworth radiometer is close enough for the purposes at hand to the one Maxwell proposes. A simulacrum account is unnecessary, the standard covering-law story will do. But this is not so. For on the covering-law theory, if Maxwell's treatment is to explain the rotation in a Woolworth radiometer, the radiometer must have a specific distribution function and that function must be nomologically linked to the conditions that obtain. But Maxwell's theory records no such laws. The conditions in these radiometers are indefinitely varied and indefinitely complex, so that a multitude of highly complicated unknown laws must be assumed to save Maxwell's explanation. I think these laws are a complete fiction. We cannot write them down. We certainly cannot construct experiments to test them. Only the covering-law model of explanation argues for their existence.

Recall Hempel's worries about bridge principles, which I discussed in the last essay. Hempel was concerned that bridge principles do not have the proper exceptionless character to ensure deductive connections between explanans and explanandum. Hempel illustrated with magnets and iron filings. But Maxwell's radiometer is as good an example. Not all radiometers that meet Maxwell's two descriptions have the distribution function Maxwell writes down; most have many other relevant features besides. This will probably continue to be true no matter how many further corrections we add. In general, as Hempel concluded, the bridge law between the medium of a radiometer and a proposed distribution can hold only *ceteris paribus*.

This, however, is a difficult position for a covering-law theorist to take. As I argued early on, in the second essay, a law that holds only in restricted circumstances can explain only in those circumstances. The bulk of the radiometers in Woolworth's are untouched by Maxwell's explanation. The idealization story with which we began supposes that each Woolworth radiometer has some distribution function true of it and that the distribution functions in question are sufficiently close to Maxwell's. In this case Maxwell's explanation for the ideal medium serves as a proxy to the proper explanations for each of the real radiometers. But these last are no explanations at all on the covering-law view unless the Book of Nature is taken up with volumes and volumes of bridge laws.

I say there are no such bridge laws, or, more cautiously, we have no reason to assume them. But without the bridge laws, the distribution functions have no explanatory power. Thus our chief motivation for believing in them vanishes, and rightly so. The distribution functions play primarily an organizing role. They cannot be seen; they cause nothing; and like many other properties of convenience, we have no idea how to apply them outside the controlled conditions of the laboratory, where real life mimics explanatory models. What is the distribution function for the molecules in this room? Or the value of the electric field vector in the region just at the tip of my pencil? These questions are queer. They are queer because they are questions with no answers. They ask about properties that only objects in models have, and not real objects in real places.

I think we are often misled by a piece of backwards reasoning here. Sometimes for a given model, it is possible to contrive (or to find) a real situation in which the principal features relevant to the phenomenology are just the features mentioned in the model, and no others. Low density helium, for example, is an almost ideal gas from the point of view of the billiard ball model of statistical mechanics. In these cases, we are inclined to think of the model as an exact replica of reality, and to attribute to the objects modelled not only the genuine properties of the model, but also the properties of convenience. By continuity, we then argue, the properties of convenience must apply to more complex cases as well. But this is just backwards. With a good many abstract theoretical properties, we have no grounds for assigning them to complex, realistic cases. By continuity, they do not apply to the ideal cases either.

Returning to models, it may help to recall a disagreement

between Mary Hesse[11] and Wilfrid Sellars.[12] Hesse's paradigm is the billiard ball model for the kinetic theory of gases. She thinks that the objects in the model (the billiard balls) and the objects modelled (the molecules of gas) share some properties and fail to share others; and she talks in terms of the positive, negative, and neutral analogies between the model and the objects modelled. Sellars disagrees. His attention is one level up. What is important for Sellars is not the sharing of properties, but the sharing of relationships among properties. I take it that our laser example would suit Sellars well. The helium-neon laser and a real triode oscillator need have no properties in common. What is relevant is that the properties each has behave in similar ways, so that both can be treated by the same van der Pol equation.

I share Sellars's stress on the relations among properties, for the point of the kind of models I am interested in is to bring the phenomenon under the equations of the theory. But Sellars and I are opposed over realism. He sees that phenomenological laws are hard to get right. If we want regular behaviour, the description of the circumstances must grow more and more complicated, the laws have less and less generality, and our statements of them will never be exceptionless. Fundamental laws, by contrast, are simple, general, and without exception. Hence for Sellars they are the basic truths of nature.

In opposition to Sellars, I have been arguing that their generality and exceptionlessness is mere appearance, appearance which arises from focusing too much on the second stage of theory entry. The fundamental equations may be true of the objects in the model, but that is because the models are constructed that way. To use the language I introduced in the last essay, when we present a model of a phenomenon, we prepare the description of the phenomenon in just the right way to make a law apply to it.

The problem for realism is the first stage of theory entry. If the models matched up one-to-one, or at least roughly

[11] Mary Hesse, *Models and Analogies in Science* (Notre Dame: University of Notre Dame Press, 1966).

[12] Wilfrid Sellars, *Philosophical Perspectives* (Reseda, California: Ridgeview Press, 1977), Ch. VIV, 'Scientific Realism or Irenic Instrumentalism'.

so, with the situations we study, the laws which govern the model could be presumed to apply to the real situations as well. But models are almost never realistic in the first sense; and I have been arguing, that is crucial to how physics works. Different incompatible models are used for different purposes; this adds, rather than detracts, from the power of the theory. We have had many examples already but let me quote one more text describing the variety of treatments available for lasers:

A number of authors have treated idealized interacting systems as models for lasers. Extensive studies have been carried out by Lax, Scully and Lamb, Haken, Sauerman, and others. Soluble models have been examined by Schwable and Therring. Several simplified dynamical models for devices of various sorts are given in the last chapter of Louisell's book.[13]

And so on.

There has been a lot of interest in models among philosophers of science lately. It will help to compare the use I make of models with other accounts. First, Redhead and Cushing: both Michael Redhead[14] and James Cushing[15] have recently done very nice investigations of models in mathematical physics, particularly in quantum mechanics and in quantum field theory. Both are primarily concerned not with Hesse's analogical models, but with what Redhead calls 'theoretical models', or incomplete theories (Cushing's model$_3$—guinea-pig or tinker toy models). Although, like me, Cushing explicitly says that models serve to 'embed' an account of the phenomena into a mathematical theory, he and Redhead concentrate on a special kind of model—a theory which is admittedly incomplete or inaccurate. I am concerned with a more general sense of the word 'model'. I think that a model—a specially prepared, usually fictional description of the system under study—is employed whenever a mathematical theory is applied to reality, and I use the word 'model' deliberately to suggest the failure of exact

[13] Klauder and Sudarshan, op. cit., pp. 234–5.
[14] Michael Redhead, 'Models in Physics', *British Journal for the Philosophy of Science* 31 (1980), pp. 145–63.
[15] James Cushing, 'Models and Methodologies in Current Theoretical High-Energy Physics', *Synthese* 50 (1982), pp. 5–101.

correspondence which simulacra share with both Hesse's analogical models and with Redhead and Cushing's theoretical models.

Secondly, the semantical view of theories: on the simulacrum account, models are essential to theory. Without them there is just abstract mathematical structure, formulae with holes in them, bearing no relation to reality. Schroedinger's equation, even coupled with principles which tell what Hamiltonians to use for square-well potentials, two-body Coulomb interactions, and the like, does not constitute a theory *of* anything. To have a theory of the ruby laser, or of bonding in a benzene molecule, one must have models for those phenomena which tie them to descriptions in the mathematical theory. In short, on the simulacrum account the model is the theory of the phenomenon. This sounds very much like the semantic view of theories, developed by Suppes[16] and Sneed[17] and van Fraassen.[18] But the emphasis is quite different. At this stage I think the formal set-theoretic apparatus would obscure rather than clarify my central points. It is easiest to see this by contrasting the points I want to make with the use to which van Fraassen puts the semantic view in *The Scientific Image*.[19]

Van Fraassen holds that we are only entitled to believe in what we can observe, and that we must remain agnostic about theoretical claims which we cannot confirm by observation. This leads him to require that only the observable substructure of models permitted by the laws of a theory should map onto the structure of the situations modelled. Only that part of a theory which is supposed to represent observable facts, and not the parts that represent theoretical facts, need be an accurate representation of how things really are.

Van Fraassen's book takes a firm stand against realism. Sellars, I have mentioned, is a profound realist. But they

---

[16] Patrick Suppes, 'Set Theoretic Structures in Science', mimeographed (Stanford: Stanford University Press, 1967).

[17] Joseph Sneed, *The Logical Structure of Mathematical Physics* (Dordrecht: Reidel, 1971).

[18] Bas van Fraassen, *The Scientific Image* (Oxford: Clarendon Press, 1980). See also 'On the Extension of Beth's Semantic Theories', *Philosophy of Science* 37 (1970), pp. 325–39.   [19] Ibid.

have in common a surprising respect for theory. Both expect that theories will get the facts right about the observable phenomena around us. For van Fraassen, the theoretical claims of a good theory need not match reality, but the claims about observables should. In a good theory, the observable substructure prescribed by the theory should match the structure of reality. This is not how I see good theories working. The observational consequences of the theory may be a rough match to what we suppose to be true, but they are generally not the best we can do. If we aim for descriptive adequacy, and do not care about the tidy organization of phenomena, we can write better phenomenological laws than those a theory can produce. This is what I have tried to show, beginning with 'Truth Doesn't Explain Much' and ending with the prepared, but inaccurate descriptions discussed in the last essay.

There is also a second important difference with van Fraassen that does not fit readily into the semantic formalism. I have talked about observational substructures in order to contrast my views with van Fraassen's. But unlike van Fraassen, I am not concerned exclusively with what can be observed. I believe in theoretical entities and in causal processes as well. The admission of theoretical entities makes my view much closer to Sellars than it earlier sounded. All sorts of unobservable things are at work in the world, and even if we want to predict only observable outcomes, we will still have to look to their unobservable causes to get the right answers.

I want to focus on the details of what actually happens in concrete situations, whether these situations involve theoretical entities or not, and how these differ from what would happen if even the best of our fundamental laws played out their consequences rigorously. In fact, the simulacrum account makes the stronger claim: it usually does not make sense to talk of the fundamental laws of nature playing out their consequences in reality. For the kind of antecedent situations that fall under the fundamental laws are generally the fictional situations of a model, prepared for the needs of the theory, and not the blousy situations of reality. I do not mean that there could never be situations to which the fundamental laws apply. That is only precluded if the theory employs properties or arrangements which are pure fictions, as I think classical statistical mechanics does. One may occur by accident, or, more likely, we may be able to construct one in a very carefully controlled experiment, but nature is generally not obliging enough to supply them freely.

Let me repeat a point I have made often before. If we are going to argue from the success of theory to the truth of theoretical laws, we had better have a large number and a wide variety of cases. A handful of careful experiments will not do; what leads to conviction is the widespread application of theory, the application to lasers, and to transistors, and to tens of thousands of other real devices. Realists need these examples, application after application, to make their case. But these examples do not have the right structure to support the realist thesis. For the laws do not literally apply to them.

The simulacrum account is not a formal account. It says that we lay out a model, and within the model we 'derive' various laws which match more or less well with bits of phenomenological behaviour. But even inside the model, derivation is not what the D-N account would have it be, and I do not have any clear alternative. This is partly because I do not know how to treat causality. The best theoretical treatments get right a significant number of phenomenological laws. But they must also tell the right causal stories. Frequently a model which is ideal for one activity is ill-suited to the other, and often, once the causal principles are understood from a simple model, they are just imported into more complex models which cover a wider variety of behaviour. For example, Richard Feynman, when he deals with light refraction in Volume II of his famous Berkeley lectures, says:

We want now to discuss the phenomenon of the refraction of light . . . by dense materials. In chapter 31 of Volume I we discussed a theory of the index of refraction, but because of our limited mathematical abilities at that time, we had to restrict ourselves to finding the index only for materials of low density, like gases. The physical principles that produced the index were, however, made clear . . . Now, however, we will find that it is very easy to treat the problem by the use of differential equations. This method obscures the physical origin of the index (as coming from the re-radiated waves interfering with the original waves), but it makes the theory for dense materials much simpler.[20]

But what is it for a theoretical treatment to 'tell' a causal story? *How* does Feynman's study of light in Volume I 'make clear' the physical principles that produce refraction? I do not have an answer. I can tell you what Feynman does in Volume I, and it will be obvious that he succeeds in extracting a causal account from his model for low density materials. But I do not have a philosophic theory about how it is done. The emphasis on getting the causal story right is new for philosophers of science; and our old theories of explanation are not well-adapted to the job. We need a theory of explanation which shows the relationship between causal processes and the fundamental laws we use to study them, and neither my simulacrum account nor the traditional covering-law account are of much help.

Causal stories are not the only problem. Even if we want to derive only pure Humean facts of association, the D-N account will not do. We have seen two ways in which it fails in earlier chapters. First, from Essay 6, the fundamental laws which start a theoretical treatment frequently get corrected during the course of the derivation. Secondly, many treatments piece together laws from different theories and from different domains, in a way that also fails to be deductive. This is the theme of Essay 3.

These are problems for any of our existing theories of explanation, and nothing I have said about simulacra helps to solve them. Simulacra do a different job. In general, nature does not prepare situations to fit the kinds of mathematical theories we hanker for. We construct both the theories and the objects to which they apply, then match them piecemeal onto real situations, deriving—sometimes with great precision—a bit of what happens, but generally not getting all the facts straight at once. The fundamental laws do not govern reality. What they govern has only the appearance of reality and the appearance is far tidier and more readily regimented than reality itself.

[20] Richard Feynman, *The Feynman Lecture on Physics*, vol. II (Reading, Mass: Addison-Wesley, 1964), p. 32.1.

Essay 9

How the Measurement Problem
is an Artefact of the Mathematics

0. INTRODUCTION

Von Neumann's classic work of 1932 set the measurement problem in quantum mechanics.[1] There are two kinds of evolution in the quantum theory, von Neumann said. The first kind is governed by Schroedinger's equation. It is continuous and deterministic. The second, called *reduction of the wave packet*, is discontinuous and indeterministic. It is described by von Neumann's projection postulate.

The terminology arises this way: classical systems have well-defined values for both position and momentum. Quantum systems, on the other hand, may have no well-defined value for either. In this case they are said to be in a *superposition* of momentum states, or of position states. Consider position. An electron will frequently behave not like a particle, but like a wave. It will seem to be spread out in space. But when a measurement is made we always find it at a single location. The measurement, we say, reduces the wave packet; the electron is projected from a wave-like state to a particle-like state.

But what is special about measurement? Most of the time systems are governed by the Schroedinger equation. Reductions occur when and only when a measurement is made. How then do measurements differ from other interactions between systems? It turns out to be very difficult to find any difference that singles out measurements uniquely. Von Neumann postulated that measurements are special because they involve a conscious observer. Eugene Wigner concurred;[2] and I think this is the only solution that succeeds in picking measurements from among all other interactions.

---

[1] John von Neumann, *Mathematical Foundations of Quantum Mechanics* (Princeton: Princeton University Press, 1955). First published in German, 1932.

[2] E. Wigner, 'Remarks on the Mind–Body Question', item 98 in I. J. Good (ed.), *The Scientist Speculates* (London: Heinemann, 1961).

It is clearly not a very satisfactory solution. The measurement problem is one of the long-standing philosophical difficulties that troubles the quantum theory.

I will argue here that the measurement problem is not a real problem. There is nothing special about measurement. Reductions of the wave packet occur in a wide variety of other circumstances as well, indeterministically and on their own, without the need for any conscious intervention. Section 1 argues that reductions occur whenever a system is prepared to be in a given microscopic state. Section 2 urges that reductions occur in other transition processes as well, notably in scattering and in decay. There is good reason for attending to these transition processes. On the conventional interpretation, which takes position probabilities as primary, quantum propositions have a peculiar logic or a peculiar probability structure or both. But transition processes, where reductions of the wave packet occur, have both a standard logic and a standard probability. They provide a non-problematic interpretation for the theory.

The proposal to develop an interpretation for quantum mechanics based on transition probabilities seems to me exactly right. But it stumbles against a new variation of the measurement problem. Two kinds of evolution are postulated. Reductions of the wave packet are no longer confined to measurements, but when do they occur? If there are two different kinds of change, there must be some feature which dictates which situations will be governed by Schroedinger's law and which by the projection postulate. The very discipline that best treats transitions suggests an answer. Quantum statistical mechanics provides detailed treatments of a variety of situations where reduction seems most plausibly to occur. This theory offers not two equations, but one. In general formulations of quantum statistical mechanics, Schroedinger's evolution and reduction of the wave packet appear as special cases of a single law of motion which governs all systems equally. Developments of this theory offer hope, I urge in Section 4, for eliminating the measurement problem and its variations entirely. The two evolutions are not different in nature; their difference is an artefact of the conventional notation.

1. IN DEFENCE OF REDUCTION OF THE WAVE PACKET

In 1975 I wrote 'Superposition and Macroscopic Observation'. Here is the problem that I set in that paper:

Macroscopic states, it appears, do not superpose. Macroscopic bodies seem to possess sharp values for all observable quantities simultaneously.

But in at least one well-known situation—that of measurement— quantum mechanics predicts a superposition. It is customary to try to reconcile macroscopic reality and quantum mechanics by reducing the superposition to a mixture. This is a program that von Neumann commenced in 1932 and to which Wigner, Groenewold, and others have contributed. Von Neumann carried out his reduction by treating measurement as a special and unique case that is not subject to the standard laws of quantum theory. Subsequent work has confirmed that a normal Schroedinger evolution cannot produce the required mixture. This is not, however, so unhappy a conclusion as is usually made out. Quantum mechanics requires a. superposition: the philosophical problem is not to replace it by a mixture, but rather to explain why we mistakenly believe that a mixture is called for.[3]

These are the words of a committed realist: the theory says that a superposition occurs; so, if we are to count the theory a good one, we had best assume it does so.

Today I hold no such view. A theory is lucky if it gets some of the results right some of the time. To insist that a really good theory would do better is to assume a simplicity and uniformity of nature that is belied by our best attempts to deal with it. In quantum mechanics in particular I think there is no hope of bringing the phenomena all under the single rule of the one Schroedinger equation. Reduction of the wave packet occurs as well, and in no systematic or uniform way. The idea that all quantum evolutions can be cast into the same abstract form is a delusion. To understand why, it is important to see how the realist programme that I defended in 1975 fails.

Quantum systems, we know, are not supposed to have

[3] N. Cartwright, 'Superposition and Macroscopic Observation', in Patrick C. Suppes (ed.), *Logic and Probability in Quantum Mechanics* (Dordrecht: D. Reidel Publishing Co.), pp. 221–34.

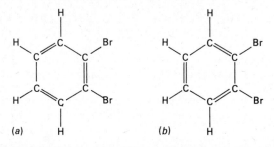

(a)   (b)

FIG 9.1.Orthodibromobenzene (*Source*:Feynman, *Lectures on Physics*.)

well-defined values for all of their variables simultaneously. A system with a precise momentum will have no position. It will behave like a wave, and be distributed across space. This is critical for the explanations that quantum mechanics gives for a variety of phenomena. The famous case of the bonding energy in benzene is a good example. This energy is much lower than would be expected if there were double bonds where they should be between the carbon atoms. The case is easier to see if we look at orthodibromobenzene—that is, benzene with two bromines substituted for two hydrogens. There are two possible structures (see Figure 9.1).

In fact all the forms in nature are the same. Do they have a single bond separating the bromines, as in 9.1*a*, or a double bond, as in 9.1*b*? Neither. The energy is in between that predicted from one bond and that predicted from two. It is as if there is a bond and a half. Linus Pauling said that each molecule resonates between the first arrangement and the second, and that is not far off the quantum answer. According to quantum mechanics the molecule is in a superposition of the two configurations. That means that it is neither in configuration 9.1*a* entirely, nor in 9.1*b* entirely, but in both simultaneously. Where then are the electrons that form the bonds? On a classical particle picture they must be located at one point or another. Quantum mechanics says that, like waves, they are smeared across the alternative locations.

This behaviour is not too disturbing in electrons. After all, they are very small and we do not have very refined intuitions about them. But it is not a correct account for macroscopic objects, whose positions, velocities, and such are narrowly fixed. How can we prevent smearing in macroscopic objects? First, we notice that smearing across one variable, like position, occurs only when an incompatible quantity, like momentum, is precisely defined. This suggests that all macroscopic observables are compatible. We can arrange this in the following way: admittedly, the real quantities that describe macroscopic systems will not all be compatible. But we do not actually observe the real value of a quantity at a time. Instead what we see is a long-time average over these values—long compared to relaxation times in the objects themselves. By a coarse-grained averaging, we can construct new quantities that are all compatible. We then claim that these new quantities, and not the originals, are the macroscopic observables that concern us.

But this is not enough. The construction ensures that it is possible for macroscopic objects to be in states with well-defined values for all macroscopic observables. It is even possible that they would always evolve from one such state into another if left to themselves. But interactions with microscopic objects bring them into superpositions. That is what happens in a measurement according to the Schroedinger equation. The electron starts out in a superposition, with the apparatus in its ground state. Together the composite of the two finishes after the measurement in a superposition, where the pointer of the apparatus has no well-defined position but is distributed across the dial.

This is why von Neumann postulated that measurement is special. In the end measurement interactions are not governed by the Schroedinger equation. After the measurement has ceased, a new kind of change occurs. The superposed state of the apparatus-plus-object reduces to one of the components of the superposition. This kind of change is called *reduction of the wave packet*, and the principle that governs it is *the projection postulate*. The process is indeterministic: nothing decides which of the final states will occur, although the probabilities for each state are antecedently determined. In an ensemble of similar measurement interactions, there will be a mix of final states. So we sometimes say that a reduction of the wave packet takes superpositions into mixtures. But it is important to remember that this is a description entirely at the level of the ensemble. Individually, reduction of the wave packet takes superpositions into components of the superposition.

There are two obvious tests to see if von Neumann is right, and both tell in favour of reduction of the wave packet. The first looks at individuals. If the pointer is indeed in a superposition after a measurement, it should somehow be distributed across space. But in fact we always observe the pointer at a definite place. The second looks at ensembles. Mixtures and superpositions generate different statistical predictions about the future behaviour of the ensemble. Collections of macroscopic systems always behave, statistically, as if they were in mixtures. In 'Superposition and Macroscopic Observation' I defend the superposition against both of these difficulties.

The first defence consists in 'breaking the eigenvalue-eigenvector link'. The discussion so far has assumed that the electron has no position *value* because it is in a superposition of position *states*. We have thus been adopting the principle: S has a value for a given observable (an eigenvalue) if and only if S is in the corresponding state (the eigenstate). To deny the inference in the direction from state to value would require serious revision of our understanding of quantum theory, which teaches that the probability in a given eigenstate for a system to exhibit the corresponding eigenvalue is one. But there is no harm in breaking the inference in the other direction. One must be careful not to run foul of various no-hidden-variable proofs, like those of Kochen and Specker[4] and of J. S. Bell.[5] But there are a variety of satisfactory ways of assigning values to systems in superpositions.[6]

Let us turn to the second difficulty. Why are physicists

[4] S. Kochen and E. P. Specker, 'The Problem of Hidden Variables in Quantum Mechanics', *Journal of Mathematics and Mechanics* 17 (1967), pp. 59–87.

[5] J. S. Bell, 'On the Problem of Hidden Variables in Quantum Mechanics', *Reviews of Modern Physics* 38 (1966), pp. 447–52.

[6] See the discussion in 'Superposition and Macroscopic Observation' and the references given there.

not much concerned with the measurement problem in practice? Here is a common answer: macroscopic objects have a very large number of degrees of freedom with randomly distributed phases. The effect of averaging over all of these degrees of freedom is to eliminate the interference terms that are characteristic of superpositions. In a very large system with uncorrelated phases the superposition will look exactly like a mixture. This argument is just like the one described in Essay 6 for deriving the exponential decay law. The atom is in interaction with the electromagnetic field. If it interacted with only one mode, or a handful of modes, it would oscillate back and forth as if it were in a superposition of excited and de-excited states. (There is a very nice discussion of this in P. C. W. Davies, *The Physics of Time Asymmetry*.)[7] In fact it is in interaction with a 'quasi-continuum' of modes, and averaging over them, as we do in the Weisskopf–Wigner approximation, eliminates the terms that represent the interference. An ensemble of such atoms evolving over time will look exactly as if it is in a mixture of excited and de-excited states, and not in a superposition.

There have been various attempts to apply these ideas more rigorously to measurement situations. The most detailed attempt that I know is in the work of Danieri, Loinger, and Prosperi.[8] This is what I describe in 'Superposition and Macroscopic Observation'. Daneri, Loinger, and Prosperi propose an abstract model of a measurement situation, and consider what happens in this model as the macroscopic apparatus comes to equilibrium after the interaction. They conclude that the superposition that the Schroedinger equation predicts will be statistically indistinguishable from the mixture predicted by the projection postulate. This is not to say that the superposition and the mixture are the same. There is no way under the Schroedinger equation

[7] P. C. W. Davies, *The Physics of Time Asymmetry* (Berkeley and Los Angeles: University of California Press, 1974, 1977), section 6.1.

[8] A. Daneri, A. Loinger, and G. M. Prosperi, 'Quantum Theory of Measurement and Ergodicity Conditions', *Nuclear Physics* 33 (1962), pp. 297–319; 'Further Remarks on the Relation Between Statistical Mecahnics and Quantum Theory of Measurement', *Il Nuovo Cimento* 44B (1966), pp. 119–28.

for the system to end up in a genuine mixture. The claim of
Daneri, Loinger, and Prosperi is weaker. Superpositions
and mixtures make different statistical predictions. But in
this case, the two will agree on predictions about all macroscopic observables. The superposition and the mixture
are different, but the difference is not one we can test
directly with our macroscopic instruments.

Formally, there is an exact analogue between measurement in the treatment of Daneri, Loinger, and Prosperi
and the case of exponential decay. In deriving the exponential decay law, we do not produce, in approximation, a
mixture as the final state. Rather we show that the superposition is indistinguishable from a mixture with respect
to a certain limited set of test variables.

Jeffrey Bub[9] and Hilary Putnam[10] have both attacked the
Daneri, Loinger, and Prosperi programme. They do not
object to the details of the measurement model. Their
objection is more fundamental. Even if the model is satisfactory, it does not solve the problem. A superposition is
still a superposition, even if it dresses up as a mixture. We
want to see a mixture at the conclusion of the measurement.
It is no credit to the theory that the superposition which it
predicts looks very much like a mixture in a variety of tests.
I now agree that a correct account will produce a mixture
for the final state. But in 1975 I thought that Bub and
Putnam had the situation backwards. Theirs is exactly the
wrong way for a realist to look at the work of Daneri,
Loinger, and Prosperi. The theory predicts a superposition,
and we should assume that the superposition occurs. The
indistinguishability proof serves to account for why we are
not led into much difficulty by the mistaken assumption
that a mixture is produced.

That was the argument I gave in 1975 in defence of the
Schroedinger equation. Now I think that we cannot avoid
admitting the projection postulate, and for a very simple
reason: we need reduction of the wave packet to account

    [9] J. Bub, 'The Daneri–Loinger–Prosperi Quantum Theory of Measurement',
Il Nuovo Cimento 57B (1968), pp. 503–20.
    [10] H. Putnam, 'A Philosopher Looks at Quantum Mechanics', in R. G. Colodny
(ed.), *Beyond the Edge of Certainty* (Englewood Cliffs: Prentice-Hall, 1965).

for the behaviour of individual systems over time. Macroscopic systems have orderly histories. A pointer that at one instant rests in a given place is not miraculously somewhere else the next. That is just what reduction of the wave packet guarantees. After the measurement the pointer is projected onto an eigenstate of position. Its behaviour will thereafter be governed by that state—and that means it will behave over time exactly as we expect.

There is no way to get this simple result from a superposition. Following the lines of the first defence above, we can assign a location to the pointer even though it is in a superposition. An instant later, we can again assign it a position. But nothing guarantees that the second position will be the time-evolved of the first. The Daneri, Loinger, and Prosperi proof, if it works, shows that the statistical distributions in a collection of pointers will be exactly those predicted by the mixture at any time we choose to look. But it does not prove that the individuals in the collection will evolve correctly across time. Individuals can change their values over time as erratically as they please, so long as the statistics on the whole collection are preserved.

This picture of objects jumping about is obviously a crazy picture. Of course individual systems do not behave like that. But the assumption that they do not is just the assumption that the wave packet is reduced. After measurement each and every individual system behaves as if it really is in one or another of the components and not as if it is in a superposition. There are in the literature a few treatments for some specific situations that attempt to replicate this behaviour theoretically but there is no good argument that this can be achieved in general. One strategy would be to patch up the realist account by adding constraints: individual values are to be assigned in just the right way to guarantee that they evolve appropriately in time. But if the superposition is never again to play a role, and in fact the systems are henceforth to behave just as if they were in the component eigenstates, to do this is just to admit reduction of the wave packet without saying so. On the other hand what role can the superposition play? None, if macroscopic objects have both well-defined values and continuous histories.

So far we have been concentrating on macroscopic objects and how to keep them out of superpositions. In fact we need to worry about microscopic objects as well. Quantum systems, we have seen, are often in superpositions. The electron is located neither on one atom nor on another in the benzene molecule. In the next section we will see another example. When an electron passes a diffraction grating, it passes through neither one or another of the openings in the grating. More like a wave, it seems to go through all at once. How then do we get microsystems out of superpositions and into pure states when we need them? To study the internal structure of protons we bombard them at high velocity with electrons. For this we need a beam, narrowly focused, of very energetic electrons; that is, we need electrons with a large and well-defined momentum. The Stanford Linear Accelerator (SLAC) is supposed to supply this need. But how does it do so?

Here is a simple model of a linear accelerator (a 'drift tube linac')—see Figure 9.2. The drift tube linac consists of two major parts: an injector and a series of drift tubes hooked up to an alternating voltage. The injector is a source of well-collimated electrons—that is, electrons which have a narrow spread in the direction of their momentum. We might get them this way: first, boil a large number of electrons off a hot wire. Then accelerate them with an electrostatic field—for example in a capacitor. Bend them with a magnet and screen all but a small range of angles. The resulting beam can be focused using a solenoid or quadrature magnets before entering the drift tube configuration for the major part of the particle's acceleration.

Arrows represent field lines at one instant of time

FIG 9.2. Model of a linear accelerator

Inside the drift tubes the electric field is always zero. In the gaps it alternates with the generator frequency. Consider now a particle of charge e that crosses the first gap at a time when the accelerating field is at its maximum. The length L of the next tube is chosen so that the particle arrives at the next gap when the field has changed sign. So again the particle experiences the maximum accelerating voltage, having already gained an energy $2eV_0$. To do this, L must be equal to $\frac{1}{2} vT$ where v is particle velocity and T the period of oscillation. Thus L increases up to the limiting value: for $v \to c$, $L \to \frac{1}{2} cT$.

The electron in the linac is like a ball rolling down a series of tubes. When it reaches the end of each tube, it drops into the tube below, gaining energy. While the ball is rolling down the second tube, the whole tube is raised to the height of the first. At the end the ball drops again. While it traverses the tube the ball maintains its energy. At every drop it gains more. After a two-mile acceleration at SLAC, the electrons will be travelling almost at the speed of light.

Imagine now that we want to bring this whole process under the Schroedinger equation. That will be very complicated, even if we use a simple diagrammatic model. Quantum systems are represented by vectors in a Hilbert space. Each system in the process is represented in its own Hilbert space—the emitting atoms, the electrostatic field, the magnetic field, the absorbing screen, and the fields in each of the separate drift tubes. When two systems interact, their joint state is represented in the product of the Hilbert spaces for the components. As we saw in the discussion of measurement, the interaction brings the composite system into a superposition in the product space for the two. When a third system is added, the state will be a superposition in the product of three spaces. Adding a fourth gives us a superposition in a four-product space, and so on. At the end of the accelerator we have one grand superposition for the electrons plus capacitor plus magnet plus field-in-the-first-tube plus field-in-the-second, and so forth.

The two accounts of what happens are incompatible. We want an electron travelling in a specified direction with a well-defined energy. But the Schroedinger equation predicts a superposition for an enormous composite, in which the electron has neither a specific direction nor a specific energy. Naïvely, we might assume that the problem is solved early in the process. A variety of directions are represented when the electrons come off the wire or out of the capacitor, but the unwanted angles are screened out at the absorbing plate. If the formal Schroedinger treatment is followed, this solution will not do. The absorbing screen is supposed to produce a state with the momentum spread across only a very small solid angle. On the Schroedinger formalism there is no way for the absorbing screen to do this job. Interaction with the screen produces another superposition in a yet bigger space. If we are to get our beam at the end of the accelerator, Schroedinger evolution must give out. Reduction of the wave packet must occur somewhere in the preparation process.

This kind of situation occurs all the time. In our laboratories we prepare thousands of different states by hundreds of different methods every day. In each case a wave packet is reduced. Measurements, then, are not the only place to look for a failure of the Schroedinger equation. Any successful preparation will do.

2. WHY TRANSITION PROBABILITIES ARE FUNDAMENTAL

Position probabilities play a privileged role in the interpretation of quantum mechanics. Consider a typical text. In the section titled 'The Interpretation of Ψ and the Conservation of Probability' Eugen Merzbacher tells us, about the wavefunction, Ψ:

Before going into mathematical details it would, however, seem wise to attempt to say precisely what Ψ is. We are in the paradoxical situation of having obtained an equation [the Schroedinger equation] which we believe is satisfied by this quantity, but of having so far given only a deliberately vague interpretation of its physical significance. We have regarded the wave function as 'a measure of the probability' of finding the particle at time t at the position $\mathbf{r}$. How can this statement be made precise?

Ψ itself obviously cannot be a probability. All hopes we might have entertained in that direction vanished when Ψ became a complex function, since probabilities are real and *positive*. In the face of this dilemma the next best guess is that the probability is proportional to $|\Psi|^2$, the square of the amplitude of the wave function ...

Of course, we were careless when we used the phrase 'probability of finding the particle *at* position $\mathbf{r}$.' Actually, all we can speak of is the probability that the particle is in a volume element d^3r which contains the point $\mathbf{r}$. Hence, we now try the interpretation that $|\Psi(\mathbf{r}, t)|^2 d^3r$ is proportional to the probability that upon a measurement of its position the particle will be found in the given volume element. The probability of finding the particle in some finite region of space is proportional to the integral of $\Psi^*\Psi$ over this region.[11]

So $|\Psi(\mathbf{r})|^2$ is a probability. That is widely acknowledged. But exactly what is $|\Psi(\mathbf{r})|^2 d^3r$ the probability *of*? What is the event space that these quantum mechanical probabilities range over? There is no trouble-free answer. Merzbacher makes the conventional proposal, that '$|\Psi(\mathbf{r}, t)|^2 d^3r$ is proportional to the probability that *upon a measurement of its position* the particle will be found in the given volume element.' But why must we refer to measurement? We should first rehearse the difficulties that beset the far simpler proposal that $|\Psi(\mathbf{r})|^2 d^3r$ represents the probability that the particle is located in the region around $\mathbf{r}$, measured or not. This answer supposes that the events which quantum probabilities describe are the real locations of quantum objects. It has a hard time dealing with situations where interference phenomena are significant—for instance, the bonding of atoms in benzene molecule, or the wave-like patterns which are observed in a diffraction experiment.

The two-slit experiment is the canonical example. A beam of electrons passes through a screen with two holes and falls onto a photographic plate. Imagine that, at the time of passing the screen, each electron is located either at slit 1 ($\mathbf{r} = s_1$) or at slit 2 ($\mathbf{r} = s_2$). Then we may reason as follows. An electron lands at a point y on the screen if and only if it lands at y and either passes through slit one or slit two, i.e.

[11] E. Merzbacher, *Quantum Mechanics*, Second Edition (Englewood Cliffs: Prentice-Hall 1970), p. 36.

$$C_1: y \equiv y \;\&\; (s_1 \lor s_2)$$

So

$$C_2: y \equiv (y \;\&\; s_1) \lor (y \;\&\; s_2)$$

It follows that

$$C_3: \text{Prob}(y) = \text{Prob}\{(y \;\&\; s_1) \lor (y \;\&\; s_2)\}$$

and, since s_1 and s_2 are disjoint

$$C_4: \text{Prob}(y) = \text{Prob}(y \;\&\; s_1) + \text{Prob}(y \;\&\; s_2)$$
$$= \text{Prob}(y/s_1)\text{Prob}(s_1) + \text{Prob}(y/s_2)\text{Prob}(s_2).$$

Now do the quantum mechanical calculation. If the source is equidistant from both slit 1 and slit 2, the probability of passing through slit 1 = probability of passing through slit 2 = $\frac{1}{2}$, and the state of the electrons at the screen will be $\Psi(y) = 1/\sqrt{2}\phi_1(y) + 1/\sqrt{2}\phi_2(y)$, where $\phi_1(y)$ is the state of an electron beam which *has passed entirely through slit 1*; $\phi_2(y)$, a beam passing *entirely through slit 2*. So, using the quantum mechanical rule

$$Q: \text{Prob}(y) = |\psi(y)|^2 = |1/\sqrt{2}\phi_1(y) + 1/\sqrt{2}\phi_2(y)|^2 =$$
$$\tfrac{1}{2}|\phi_1(y)|^2 + \tfrac{1}{2}|\phi_2(y)|^2 + \tfrac{1}{2}\phi_1(y)\phi_2^*(y) + \tfrac{1}{2}\phi_1^*(y)\phi_2(y).$$

But identifying

$$Q_1: |\phi_1(y)|^2 = \text{Prob}(y/s_1)$$

$$Q_2: \tfrac{1}{2} = \text{Prob}(s_1),$$

and similarly for s_2, we see that the classical calculation C for the probability of landing at y, and the quantum mechanical calculation Q do not give the same results. They differ by the interference terms $\tfrac{1}{2}\phi_1(y)\phi_2^*(y) + \tfrac{1}{2}\phi_1^*(y)\phi_2(y)$. The C calculation is the one that supposes that the electrons have a definite location at the screen and its consequences are not borne out in experiment.

There are various ways to avoid the conclusion C_4, well-rehearsed by now. The first insists that quantum mechanical propositions have a funny logic. In particular they do not obey the distributive law. This blocks the argument at the move between C_1 and C_2. This solution was for a while persuasively urged by Hiliary Putnam.[12] Putnam has now given it up because of various reflections on the meaning and use of the connective *or*; but I am most impressed by the objections of Michael Gardner[13] and Peter Gibbins[14] that the usual quantum logics do not in fact give the right results for the two-slit experiment.

The second well-known place to block the C derivation is between the third and the fourth steps. The first attempt maintained that quantum propositions have a peculiar logic; this attempt insists that they have a peculiar probability structure. It is a well-known fact about quantum mechanics that the theory does not define joint probabilities for non-commuting (i.e. incompatible) quantities. Position and momentum are the conventional examples. But the different positions of a single system at two different times are also incompatible quantities, and so their joint probabilities are also not defined. (The incompatibility of 'r at t' and 'r at t'' is often proved on the way to deriving the theorem of Ehrenfest that, in the mean, quantum mechanical processes obey the classical equations of motion.) So Prob(y & s_1) and Prob(y & s_2) from C_4 do not exist. If Prob(y) is to be calculated it must be calculated in some different way — notably, as in the Q derivation.

What does it mean in a derivation like this to deny a joint probability? After all, by the time of C_3 we have gone so far as to admit that each electron, individually, has a well-defined position at both the screen and at the plate. What goes wrong when we try to assign a probability to this conjunctive event? Operationally, the failure of joint distributions must show up like this. We may imagine charting finite histograms for the joint values of 'r at t' and 'r at t'', but the histograms forever bounce about and never converge to a single curve. There is by hypothesis some joint frequency

[12] H. Putnam, 'Is Logic Empirical?' in R. Cohen and M. Wartofsky (eds), *Boston Studies in the Philosophy of Science*, 5 (Dordrecht: D. Reidel, 1968).
[13] Michael R. Gardner, 'Is Quantum Logic Really Logic?' *Philosophy of Science* 38 (1971), pp. 508–29.
[14] P. Gibbins, forthcoming in *Foundations of Physics*.

in every finite collection, but these frequencies do not approach any limit as the collections grow large. This may not seem surprising in a completely chaotic universe, but it is very surprising here where the marginal probabilities are perfectly well defined: in the very same collections, look at the frequencies for 'r at t' or 'r' at t' separately. These frequencies can always be obtained by summing over the joint frequencies: e.g., in any collection the frequency (y) = frequency $(y \; \& \; s_1)$ + frequency $(y \; \& \; s_2)$. As the collections grow larger, the sum approaches a limit, but neither of the terms in the sum do. This is peculiar probabilistic behaviour for a physical system. (Though since, as I remarked in the last essay, I am suspicious even of the corresponding probabilities in classical statistical mechanics, my own view is that it is considerably less peculiar than quantum logic.)

Before proceeding, it is important to notice that both attempts to begin with C_1 and to block the inference to C_4 depend on incompatibility. It is the incompatibility of location-when-passing-the-screen and location-at-the-plate that stops the inference from C_1 to C_2. In conventional quantum logics, the distributive law holds if all the propositions are compatible. Similarly, the theory will always give a joint distribution for any two quantities that commute.

We began with the question, what are quantum probabilities probabilities *of*? Practitioners of quantum theory have been reluctant to adopt either non-standard logics or non-standard probabilities. They have rejected the first proposal altogether. Quantum probabilities are not probabilities that the system is 'located at r', but rather, as Merzbacher says, that it 'will be found at r in a position measurement'. This answer is no more trouble-free than the first. It supposes that the electron passes through neither one slit nor the other when we are not looking. When we do look, there, suddenly, it is, either at the top slit or at the bottom. What is special about looking that causes objects to be places where they would not have been otherwise? This is just another version of the notorious measurement problem, which we first discussed in the last section.

I find neither of these two conventional answers very

satisfactory, and I propose a more radical alternative. I want to eliminate position probabilities altogether, and along with them the probabilities for all the classic dynamic quantities. The only real probabilities in quantum mechanics, I maintain, are transition probabilities. In some circumstances a quantum system will make a transition from one state to another. Indeterministically and irreversibly, without the intervention of any external observer, a system can change its state: the quantum number of the new state will be different and a quantum of some conserved quantity—energy, or momentum, or angular momentum, or possibly even strangeness—will be emitted or absorbed. When such a situation occurs, the probabilities for these transitions can be computed; it is these probabilities that serve to interpret quantum mechanics.

I shall illustrate with a couple of examples: the first, exponential decay; and the second, the scattering of a moving particle by a stationary target. Whether in an atom when an outer shell electron changes orbit and a photon is given off, or in a nucleus resulting in α, β, or γ radiation, decay is the classic indeterministic process. Consider a collection of non-interacting atoms, each in an excited state. The number of atoms that continue to be in the excited state decreases exponentially in time. One by one the atoms decay, and when an atom decays, something concrete happens—a photon is given off in a specific direction, the energy of the electromagnetic field is increased, and its polarization is affected. Which atom will decay, or when, is completely indeterministic. But the probability to decay is fixed, and this transition probability is the probability on which I want to focus.

The second example is from scattering theory—a theory of critical importance in high energy physics, where fundamental particles are studied almost entirely through their behaviour in collisions. Consider a beam which traverses a very long tube of very small diameter and then collides with a target of massive particles and is scattered. If we ring the target with detectors, we will find that many of the incoming particles miss the target and continue in the forward direction. But a number of others will be scattered through a variety of angles. If the scattering is elastic, the incoming particles will change only their direction; in inelastic scattering the moving particles will exchange both momentum direction and energy with the target. The transition probabilities in this case are the probabilities for a particle whose momentum was initially in the forward direction to be travelling after scattering in another specific direction, and with a well-defined energy.

The transition probabilities that occur in scattering have a venerable history: they are the first to have been introduced into quantum mechanics. Max Born's preliminary paper of 1926 is generally agreed to provide the original proposal for interpreting the wave function probabilistically. The abstract of that paper reads: 'By means of an investigation of collision processes the viewpoint is developed that quantum mechanics in Schrödinger's form can be used to describe not only stationary states but also quantum jumps.'[15] Born treats a collision between a moving electron and an atom. In the middle of the paper he says

If one now wishes to understand this result in corpuscular terms, then only one interpretation is possible: $\Phi_{n_\tau m(\alpha,\beta,\gamma)}$ defines the probability[1] that the electron coming from the z-direction will be projected in the direction defined by (α, β, γ) (and with phase change δ), where its energy I has increased by a quantum $h\nu^{\circ}_{nm}$ at the expense of the atomic energy.[16]

The footnote is famous:

1. Remark added in proof corrections: More precise analysis shows that the probability is proportional *to the square of* $\Phi_{n_\tau m(\alpha,\beta,\gamma)}$.[17]

So probability entered quantum mecahnics not as a position probability or a momentum probability, but rather as the transition probability 'that the electron coming from the z-direction will be projected in the direction defined by $(\alpha. \beta, \gamma)$'.

I am urging that we give up position probabilities,

[15] M. Born, 'Zur Quantenmechanik der Stossvorgange', *Zeitschrift fur Physik* 37 (1926), p. 863. Quotations taken from an English translation prepared by Martin Curd for a seminar in the Department of History and Philosophy of Science at the University of Pittsburgh, Spring 1978.

[16] Ibid., pp. 865–6.

[17] Ibid., p. 865, italics added.

momentum probabilities, and the like, and concentrate instead on transition probabilities. The advantage of transition probabilities is that they have a classical structure, and the event space over which they range has an ordinary Boolean logic. To understand why, we need for a moment to look at the formal treatment of transitions. Transitions occur in situations where the total Hamiltonian H is naturally decomposed into two parts—H_0, the 'free' Hamiltonian in the situation, and V, a disturbing potential: $H = H_0 + V$. The system is assumed to begin in an eigenstate of H_0. Supposing, as is the case in examples of transitions, that H_0 does not commute with V and hence does not commute with H, the eigenvalue of H_0 will not be conserved. Formally, the initial eigenstate of H_0 will evolve into a new state which will be a superposition of H_0 eigenstates: $\psi(t) = \Sigma c_i(t)\phi_i$ where $H_0 = \Sigma \alpha_i |\phi_i\rangle\langle\phi_i|$. The transition probabilities are given by the $|c_i(t)|^2$. In the old language, they are the probabilities for a system initially in a given ϕ_0 to 'be' or 'be found' later, at t, in some one or another of the eigenstates ϕ_i. The probabilities and the event space they range over are classical just because we look at no incompatible observables. It is as if we were dealing with the possessed values for H_0. All observables we can generate from H_0 will be mutually compatible, and the logic of compatible propositions is classical. So too is the probability.

The primary consideration that has made philosophers favour quantum logic is the drive towards realism (though Alan Stairs[18] lays out other, more enduring grounds). They want to find some way to ensure that a quantum system will possess values for all the classic dynamic quantities—position, momentum, energy, and the like. But this motivation is ill founded. If we want to know what properties are real in a theory, we need to look at what properties play a causal role in the stories the theory tells about the world. This is a point I have been urging throughout this book, and it makes a crucial difference here. Judged by this criterion, the classical dynamic quantities fare badly. The static values

[18] A. Stairs, 'Quantum Logic and Interpretation', talk give at the Annual Meeting of the Society for Exact Philosophy, Tempe, Arizona, 26 February 1982 (unpublished).

of dynamic variables have no effects; it is only when systems exchange energy or momentum, or some other conserved quantity, that anything happens in quantum mechanics. For example, knowing the position of a bound electron tells us nothing about is future behaviour, and its being located wherever it is brings about nothing. But if the electron jumps orbits—that is, if the atom makes an energy transition—this event has effects that show up in spectroscopic lines, in the dissolution of chemical bonds, in the formation of ions, and so on. This is true even for measurements of position. The detector does not respond to the mere presence of the system but is activated only when an exchange of energy occurs between the two. We need to consider an example that shows how transitions play a causal role in quantum mechanics which the conventional dynamic quantities do not. This seems a good one to use.

Henry Margenau[19] has long urged that all quantum measurements are ultimately measures of position. But position measurements themselves are basically records of scattering interactions. Position measurements occur when the particle whose position is to be measured scatters from a detector. The scattering is inelastic: energy is not conserved in the particle and the detector is activated by the energy that the particle gives up as it scatters. In the common position-measuring devices—cloud chambers, scintillation counters, and photographic plates—the relevant scattering interaction is the same. The particle scatters from a target in the detector, and the energy that is given up by the particle causes the detector to ionize. The devices differ in how they collect or register the ions; but in each case the presence of the particle is registered only if the appropriate ionization interaction occurs. So, at best, the probability of a count in the detecting device is equal to the probability of the designated ionization interaction.

Matters may be much worse. General background radiation may produce ionization when no particle is present. Conversely, the ion-collecting procedure may be inefficient and

[19] H. Margenau, 'Critical Points in Modern Physical Theory', *Philosophy of Science* 4 (1937), pp. 352–6; 'Philosophical Problems Concerning the Meaning of Measurement in Physics', *Philosophy of Science* 25 (1958), pp. 23 ff.

ions that are produced by the scattered particle may fail to register. Photographic emulsions are highly efficient in this sense; efficiencies in some cases are greater than 98 per cent. But other devices are not so good. In principle it is possible to correct for these effects in calculating probabilities. For simplicity I shall consider only devices that are ideally efficient—that is, I shall assume that all and only ions that are produced by the scattered particle are collected and counted.

I have urged that a real detector cannot respond to the mere presence of a particle, but will react only if the particle transfers energy to it. If the amplification process is maximally efficient so that the counter registers just in case the appropriate energy is transferred, then a particle will register in the detector if and only if the appropriate energy interchange occurs. This could create a serious consistency problem for the theory: a particle will be found at r in a position measurement just in case a detector located at r undergoes some specific energy interaction with the particle. As we have been discussing, the probability of the first event is supposed to be $|\psi(\mathbf{r})|^2 d^3r$. But the probability of the second is calculated in a quite different way, by the use of the methods of scattering theory. Quantum mechanics is consistent only if the two probabilities are equal, or approximately so. Otherwise $|\psi(\mathbf{r})|^2$ will not give the probability that the system will be found at r in a real physical measurement.

In fact this is stronger than necessary, for we are not interested in the absolute values of the probabilities, but merely in their relative values. Consider, for example, the photographic plate, which is the device best suited for determining densities of particles at fairly well-defined positions. In a photographic plate we are not concerned that the number of spots on the plate should record the actual number of particles, but rather that the pattern on the plate should reflect the distribution of particles. This requires establishing not an equality, but merely a proportionality:

$$\frac{\text{Probability of ionization in a target at } r}{\text{Probability of ionization in a target at } r'} = \frac{|\psi(r)|^2}{|\psi(r')|^2}. \qquad (9.1)$$

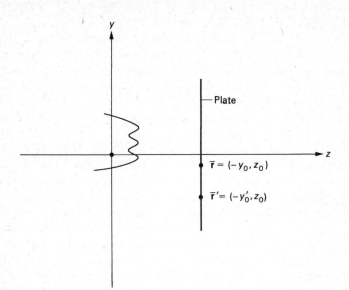

FIG 9.3. Origin located at source of wave

For a general justification, (9.1) should be true for any state function ψ for which the problem is well defined. For simplicity I will treat a two-dimensional example, in which a row of detectors is arrayed in a line perpendicular to the z axis. We may think of the detectors as the active elements in a photographic plate. In this case $\psi(r, t)$ should be arbitrary in shape, except that at $t = 0$ it will be located entirely to the left of the plate (see Figure 9.3). It is easiest to establish (9.1) not in its immediate form, but rather by inverting to get

$$\frac{\text{Probability of ionization in a target at } r}{|\psi(r)|^2}$$

$$= \frac{\text{Probability of ionization in a target at } r'}{|\psi(r')|^2}. \qquad (9.2)$$

Thus, the aim is to show that the ratio in (9.2) is a constant independent of r. A shift of coordinates will help. We think of fixing the centre of the wave and varying the location of

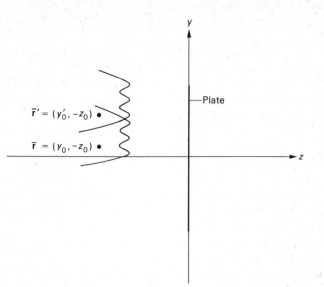

FIG 9.4. Origin located at detector

the detector across the plate (Figure 9.3). But the same effect is achieved by fixing the location of the detector and varying the centre of the wave (Figure 9.4). Looked at in this way, the consistency result can be seen as a trivial consequence of a fundamental theorem of scattering theory. This theorem states that the scattering cross section, both total and differential, is a constant independent of the shape or location of the incoming wave packet. The total scattering cross section is essentially the ratio of the total probability for the particle to be scattered, divided by the probability for crossing the detector. Roughly, the theorem assures that the probability of scattering from the detector for a wave centred at r_0, divided by the probability for 'being at' the detector, is a constant independent of r_0. This is just what is required by (9.2).

There is one difficulty. Standard textbook proofs of this theorem do not establish the result for arbitrary waves, but only for wave packets that have a narrow spread in the *direction of the momentum*. This is not enough to ensure consistency. In a previous paper I have calculated (9.2) directly and shown it to hold for arbitrary initial states.[20]

I have been urging that the interpretation of quantum mechanics should be seen entirely in terms of transition probabilities. Where no transitions occur, ψ must remain uninterpreted, or have only a subjunctive interpretation: if the system were subject to a perturbing potential of the right sort, the probability for a transition into state——would be ——. This makes the theory considerably less picturesque. We think in terms of where our microsystems are located, and all our instincts lead us to treat these locations as real. It is a difficult matter to give up these instincts; but we know from cases like the two-slit experiment that we must, if our talk is to be coherent.

We should consider one more example in detail, to see exactly how much of our reasoning in quantum mechanics relies on picturing the positions of quantum systems realistically, and what further steps we must take if we are going to reject this picture. Dipole radiation is one of the clearest examples where position seems to count. Recall from Essay 7 that the atoms in a laser behave very much like classical electron oscillators. I will use the treatment in Sargent, Scully, and Lamb to make it easy for the reader to follow, but the basic approach is quite old. Dipole radiation is one of the very first situations to which Schroedinger applied his new wave mechanics. Sargent, Scully and Lamb tell us:

It is supposed that quantum electrons in atoms effectively behave like charges subject to the forced, damped oscillation of an applied electromagnetic field. It is not [at] all obvious that bound electrons behave in this way. Nevertheless, the average charge distribution does oscillate, as can be understood by the following simple argument. We know the probability density for the electron at any time, for this is given by the electron wave function as $\psi^*(\mathbf{r}, t)\psi(\mathbf{r}, t)$. Hence, the effective charge density is

$$e\psi^*(\mathbf{r}, t)\psi(\mathbf{r}, t).$$

[20] N. Cartwright, 'Measuring Position Probabilities', in P. Suppes (ed.), *Studies in the Foundations of Quantum Mechanics* (East Lansing: Philosophy of Science Association, 1980), pp. 109–18.

For example, consider a hydrogen atom initially in its ground state with [a] spherical distribution . . . Here the average electron charge is concentrated at the center of the sphere. Application of an electric field forces this distribution to shift with respect to the positively charged nucleus . . . Subsequent removal of the field then causes the charged sphere to oscillate back and forth across the nucleus because of Coulomb attraction. This oscillating dipole acts something like a charge on a spring.[21]

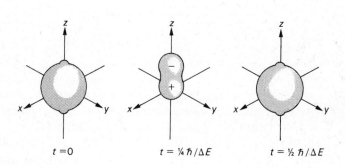

$t = 0$ $t = \frac{1}{4}\hbar/\Delta E$ $t = \frac{1}{2}\hbar/\Delta E$

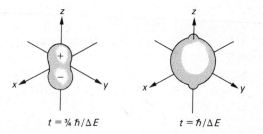

$t = \frac{3}{4}\hbar/\Delta E$ $t = \hbar/\Delta E$

FIG 9.5. Oscillating charge distribution in a hydrogen atom (*Source*: Siegman, *Lasers*).

Consider as a concrete example a hydrogen atom in its ground state (the 1s state, which I shall designate by $U_a(\mathbf{r})$). If the atom is subject to an external electric field, it will evolve into a superposition of excited and de-excited states,

[21] M. Sargent, M. Scully, and W. Lamb, *Laser Physics* (Reading: Addison-Wesley, 1974), p. 31.

as we have seen before. (Here I will take the excited state to be the $2p$ state, with $m = 0$, designated by $U_b(\mathbf{r})$.) So at t the state of the hydrogen atom in an electric field is

$$\psi(\mathbf{r}, t) = C_a(t)\exp(-i\omega_a t)U_a(\mathbf{r}) + C_b(t)\exp(-i\omega_b t)U_b(\mathbf{r}).$$

If we chart the charge density $e\,|\psi(\mathbf{r}, t)|^2$ at intervals $\Delta t = \frac{1}{4}(\hbar/\Delta E)$ apart, we see from Figure 9.5 that the charge distribution moves in space like a linear dipole, and hence has a dipole moment.[22] The dipole moment for the atom is given by

$$\langle e\mathbf{r} \rangle = pC_a C_b^*\exp\{-i(\omega_a - \omega_b)t\} +$$
$$+ pC_a^* C_b \exp\{+i(\omega_a - \omega_b)t\}.$$

The brackets, $\langle e\mathbf{r}\rangle$, on the left indicate that we are taking the quantum mechanical expectation. As Sargent, Scully, and Lamb say, 'The expectation value of the electric dipole moment $\langle e\mathbf{r}\rangle$ is given by the average value of $e\mathbf{r}$ with respect to this probability density [the density $\psi^*(\mathbf{r})\psi(\mathbf{r})$]'.[23] The location of the electron is thus given a highly realistic treatment here: the dipole moment at a time t is characterized in terms of the *average position* of the electron at t.

Classically an oscillating dipole will radiate energy. The quantum analogue of this dipole radiation is central to the Lamb theory of the laser. In a laser an external field applied to the cavity induces a dipole moment in the atoms in the lasing medium. The calculation is more complicated, but the idea is the same as in the case of the hydrogen atom that we have just looked at. Summing each of the microscopic dipole moments gives the macroscopic moment, or polarization, of the medium. This in turn acts as a source in Maxwell's equations. A condition of self-consistency then requires that the assumed field must equal the reaction field. Setting the two equal, we can derive a full set of equations to describe oscillation in a laser.

I have two comments about the realistic use of electron

---

[22] Figure 9.5 is taken from an unpublished draft of a forthcoming second edition of A. E. Siegman, *An Introduction to Lasers and Masers*, where it appears as figure 2-16 on p. 2-49.

[23] Op. cit. footnote 20, p. 3.

position in the Lamb theory. The first is fairly instrumentalist in tone. The principal equations we derive in this kind
of treatment are *rate equations* either for the photons or for
the atoms. These are equations for the time rate of change
of the number of photons or for the occupation numbers
of the various atomic levels: that is, they are equations for
transition probabilities. As with scattering or simple decay,
these probabilities are completely classical. The equations
are just the ones that should result if the atoms genuinely
make transitions from one state to another. Louisell, for
instance, in his discussion of Lamb's treatment derives an
equation for the occupation number N_a, of the excited state:

$$\text{L.8.3.21: } dN_a/dt = R_a(t) - \Gamma_a N_a(t) +$$
$$+ i/\hbar[P_n(t)E^*(t) - P_n^*(t)E(t)].$$

He tells us:

The physical meaning of these equations should be quite clear. Equation (8.3.21) gives the net rate of change at which atoms are entering
and leaving state $|a\rangle$. The R_a term gives the rate at which atoms are
being 'pumped' into level $|a\rangle$. The $-\Gamma_a N_a$ term represents the inco-
herent decay of atoms from level $|a\rangle$ to lower levels. We could also
add a term $+ W_{ab}N_b$ to represent incoherent transitions from $|b\rangle$ to
$|a\rangle$, but we omit this for simplicity. The Γ_a^{-1} is the lifetime of the atom
in level $|a\rangle$ in the absence of [a] driving field. These first terms are
incoherent since they contain no phase information. . .
    The last term $i(PE^* - P^*E)$ represents the net induced population
change in level $|a\rangle$ due to the presence of a driving field.[24]

What is important to notice is that the change in the number
of excited atoms is just equal to the number of atoms going
into the excited state minus the number going out. There
are no terms reflecting interference between the excited
and the de-excited states.
    Recall that in Essay 6 we showed that the exponential
decay law can be drived by a Markov approximation, which
assumes that there is no correlation among the reservoir
variables. Exactly the same device is used to eliminate the

---

[24] W. H. Louisell, *Quantum Statistical Properties of Radiation* (New York:
John Wiley & Sons, 1973), p. 459.

non-classical interference terms here. For example, in deriving the photon rate equation, H. Haken remarks:

We shall neglect all terms with $b_{\lambda'}^* b_\lambda$ ($\lambda \neq \lambda'$). This can be justified as follows. . . The phases of the b^*'s fluctuate. As long as no phase locking occurs the phase fluctuations of the different b's are uncorrelated. The mixed terms on the right hand side of 134 [i.e. the interference terms in the photon rate equation] vanish if an average over the phases is taken.[25]

The classical character of the rate equations leaves us in a peculiar position. On the one hand, the superposition of excited and de-excited states is essential to the theoretical account. There is no dipole moment without the superposition—the charge density $e|\psi(\mathbf{r})|^2$ does not oscillate in space in either the excited state alone or in the de-excited state alone. Without the dipole moments, there is no macroscopic polarization produced by the medium, and hence no grounds for the self-consistency equations with which Lamb's derivations begin. On the other hand, the classical character of the rate equations suggests that the atoms genuinely change their states. If so, this is just like the picture I have sketched for simple decay cases: the atom makes a transition, Schroedinger evolution stops and reduction of the wave packet takes over. The very formalism that allows us to predict that this will occur does not apply throughout the entire process.

This leads me to suppose that the whole account is an explanatory fiction, including the oscillating dipoles, whose role is merely to motivate us to write down the correct starting equation. The account is best seen as a simulacrum explanation. The less radical response is to notice that Lamb and others describe $e|\psi(\mathbf{r})|^2$ as 'the charge distribution'. In this very simple case, we need not see it as a probability distribution at all, but as a genuine charge density in space. This of course will not do as a general interpretation of $\psi^*\psi$. (Generally ψ must be taken as a function in a phase space. Where more than one electron is involved, ψ will be a function of the position of all of the electrons; and in

[25] H. Haken, 'The Semi-classical and Quantum Theory of the Laser', in S. M. Kay and A. Maitland (eds), *Quantum Optics* (London: Academic Press, 1970), p. 226.

general $\psi^*\psi$ in the phase space will not reduce to a simple distribution in real space for the charges involved.) But for any case where we want to claim that the dipole moment arises from a genuine oscillation, this must be possible. Otherwise the correct way to think about the process is not in terms of probabilities at all, but rather this way: the field produced by the atoms contributes to the macroscopic polarization. $\langle er \rangle$ is the quantity in the atoms responsible for the micro-polarization it produces. We learn to calculate this quantity by taking $e|\psi(r)|^2$, and this calculation is formally analogous to computing a moment or an average. But the analogy is purely formal. $e|\psi(r)|^2$ is a quantity which gives rise to polarization in the field; it has no probabilistic interpretion. This way of viewing 'expectations' is not special to this case. Whenever an expectation is given a real physical role, it must be stripped of its probabilistic interpretation. Otherwise, how could it do its job? $\langle er \rangle$ certainly is not the average of actual possessed positions, since electrons in general do not possess positions. The conventional alternative would have it be the average of positions which would be found on measurement. But that average can hardly contribute to the macroscopic polarization in the unmeasured cavity.

In this section I have argued that two independent reasons suggest interpreting quantum mechanics by transition probabilities and not by position probabilities or probabilities for other values of classic dynamic quantities. First, transition events play a causal role in the theory which is not matched by the actual positions, momenta, etc. of quantum systems. The second is that transition probabilities are classical, and so too are their event spaces, so there is no need for either a special logic or a special probability. Both of these arguments suppose that transitions are events which actually occur: sometime, indeterministically, the system does indeed change its state. This is the view that Max Born defended throughout his life. We find a modern statement of it in the text of David Bohm:

We conclude that $|C_m|^2$ yields the probability that a transition has taken place from the sth to the mth eigenstate of H_0, since the time $t = t_0$. Even though the C_m's change continuously at a rate determined by Schrödinger's equation and by the boundary conditions at $t = t_0$, the system actually undergoes a discontinuous and indivisible transition from one state to the other. The existence of this transition could be demonstrated, for example, if the perturbing potential were turned off a short time after $t = t_0$, while the C_m's were still very small. If this experiment were done many times in succession, it would be found that the system was always left in some eigenstate of H_0. In the overwhelming majority of cases, the system would be left in its original state, but in a number of cases, proportional to $|C_m|^2$, the system would be left in the mth state. Thus, the perturbing potential must be thought of as causing indivisible transitions to other eigenstates of H_0.[26]

But this view is not uncontroversial. To see the alternative views, consider radioactive decay. There are two ways to look at it, one suggested by the old quantum theory, the other by the new quantum theory. The story told by the old quantum theory is just the one with which most of us are familiar and which I adopt, following Bohm. First, radioactive decay is indeterministic; second, the activity decreases exponentially in time; and third, it produces a chemical change in the radioactive elements. Henri Becquerel reported the first observations of radioactivity from uranium in a series of three papers in 1886. Marie Curie did a systematic study of uranium and thorium beginning in 1898, for which she and Pierre Curie shared a Nobel prize with Becquerel. But it was not until the work of Rutherford and Soddy in 1902 that these three important facts about radioactivity were recognized. The first and second facts come together. The probability for the material to remain in its excited state decreases exponentially in time, and no external influence can affect this probability, either to increase it or to decrease it. Rutherford and Soddy report:

It will be shown later that the radioactivity of the emanation produced by thorium compounds decays geometrically with the time under all conditions, and is not affected by the most drastic chemical and physical treatment. The same has been shown by one of us to hold for the excited radioactivity produced by the thorium emanation.

[26] D. Bohm, *Quantum Theory* (Englewood Cliffs: Prenctice-Hall, 1951).

This decays at the same rate whether on the wire on which it is originally deposited, or in the solution of hydrochloric or nitric acid. The excited radioactivity produced by the radium emanation appears analogous.[27]

The third fact is the one of relevance to us. From Rutherford and Soddy's introduction: 'Radioactivity is shown to be accompanied by chemical changes in which new types of matter are being continuously produced.'[28] During decay, uranium 238 is transformed into thorium 234. When the alpha particle is emitted, the state of the material changes. This is exactly analogous to Einstein's treatment of atomic decay in his derivation of the Plank law of black body radiation (though Einstein's real feeling about this situation seems to have been far more ambiguous). Concerning what we now call spontaneous emission he says, 'This is a transition from the state Z_m into the state Z_n together with emission of the radiation energy $E_m - E_n$. This transition takes place without any external influence. One can scarcely avoid thinking of it as like a kind of radioactive reaction.'[29] Bohr too has the same picture when he first quantizes the energy levels of the atom. Only certain orbits are allowed to the electrons. In radiation the electron changes from one fixed orbit to another, emitting a quantum of light energy. In discussing the spectrum of hydrogen Bohr says, 'During the emission of the radiation the system may be regarded as passing from one state to another.'[30] On the old-quantum-theory picture the time of decay is undetermined. But when decay occurs, a photon, or an alpha particle or a beta particle is given off, and the emitting material changes its state.

Contrast this with the new-quantum-theory story. This is the story we read from the formalism of the developed mathematical theory. On this story nothing happens. In

[27] E. Rutherford and F. Soddy, 'The Cause and Nature of Radioactivity', *Philosophical Magazine* 4 (1902), p. 387.

[28] Ibid., p. 371.

[29] A. Einstein, 'Strahlungs-Emission und -Absorbtion nach der Quantentheorie', *Deutsche Physicalische Gesellschaft, Verhandlungen* 18 (1916), p. 321. Reference and translation provided by Brent Mundy, Stanford University.

[30] N. Bohr, 'On the Spectrum of Hydrogen', address before Physical Society Copenhagen, 1913, p. 11, quoted in H. A. Boorse and L. Motz (eds), *The World of the Atom* (New York: Basic Books, 1966), p. 747.

atomic decay the atom begins in its excited state and the field has no photons in it. Over time the composite atom-plus-field evolves continuously under the Schroedinger equation into a superposition. In one component of the superposition the atom is still in the excited state and there are no photons present; in the other, the atom is de-excited and the field contains one photon of the appropriate frequency. The atom is neither in its outer orbit nor in its inner orbit, and the photon is neither there in the field travelling away from the atom with the speed of light, nor absent. Over time the probability to 'be found' in the state with an excited atom and no photons decays exponentially. In the limit as t → ∞, the probability goes to zero. But only as t → ∞! On the new-quantum-theory story, never, at any finite time, does an atom emit radiation. Contrary to Bohr's picture, the system may *never* be regarded 'as passing from one state to another'.

The situation with scattering is no better. A particle with a fixed direction and a fixed energy bombards a target and is scattered. The state of the scattered particle is represented by an outgoing spherical wave (see Figure 9.6 in the Appendix). After scattering the particle travels in no fixed direction; its outgoing state is a superposition of momentum states in all directions. We may circle the target with a ring of detectors. But as we saw in discussing the problem of preparation, this is of no help. If we look at the detectors, we will find the particle recorded at one and only one of them. *We* are then cast into a superposition; each component self sees the count at a different detector. It is no wonder that von Neumann says that here at least reduction of the wave packet occurs. But his reduction comes too late. Even without the detectors, the particle must be travelling one way or another far away from the target.

So on my view, as in the old quantum theory, reduction of the wave packet occurs in a variety of situations, and independent of measurement. Since I have said that superpositions and mixtures make different statistical predictions, this claim ought to be subject to test. But direct statistical test will not be easy. For example, to distinguish the two in the case of atomic decay, we would have to do a correlation experiment on both the atom and its associated photon, and we would have to measure some observable which did not commute with either the energy levels of the atom or the modes of the perturbed field. (This is laid out formally in 'Superposition and Macroscopic Observation'.) But these kinds of measurements are generally inaccessible to us. That is what the work of Daneri, Loinger, and Prosperi exploits. Still, with ingenuity, we might be able to expose the effects of interference in some more subtle way.

P. H. Eberhard has proposed some experiments which attempt to do so.[31] Vandana Shiva proposes a test as well.[32] Not all tests for interferences will be relevant, of course, for reductions occur, but not all of the time. Otherwise there would be no interference pattern on the screen in the two-slit experiment, and the bonding energy of benzene would be different. But there is one experiment that Eberhard proposes which I would count as crucial. This is a test to look for reduction of the wave packet in one of the very cases we have been considering—in scattering. I will discuss Eberhard's experiment in the Appendix.

3. HOW THE MEASUREMENT PROBLEM IS AN ARTEFACT OF THE MATHEMATICS

Von Neumann claimed that reduction of the wave packet occurs when a measurement is made. But it also occurs when a quantum system is prepared in an eigenstate, when one particle scatters from another, when a radioactive nucleus disintegrates, and in a large number of other transition processes as well. That is the lesson of the last two sections. Reductions of the wave packet go on all of the time, in a wide variety of circumstances. There is nothing peculiar about measurement, and there is no special role for consciousness in quantum mechanics.

This is a step forward. The measurement problem has

---

[31] P. H. Eberhard, 'Should Unitarity Be Tested Experimentally?', *CERN Report* No. CERN 72–1 (1972, unpublished).

[32] V. Shiva, 'Are Quantum Mechanical Transition Probabilities Classical? A Critique of Cartwright's Interpretation of Quantum Theory', *Synthese* 44 (1980), pp. 501–8.

disappeared. But it seems that another has replaced it. Two kinds of evolution are still postulated: Schroedinger evolution and reduction of the wave packet. The latter is not restricted to measurement type situations, but when does it occur? What features determine when a system will evolve in accord with the Schroedinger equation and when its wave packet will be reduced? I call this problem *the characterization problem*. I am going to argue that it is no real problem: it arises because we mistakenly take the mathematical formulation of the theory too seriously. But first we should look at a more conventional kind of answer.

There is one solution to the characterization problem that is suggested by the formal Schroedinger attempts to minimize interference terms: reduction of the wave packet occurs when and only when the system in question interacts with another *which has a very large number of independent degrees of freedom*. Recall the derivations of exponential decay discussed in Essay 6. In the Weisskopf–Wigner treatment we assume that the atom couples to a 'quasi-continuum' of modes of the electromagnetic field. If instead it coupled to only one, or a few, the probability would not decay in time, but would oscillate back and forth forever between the excited and de-excited states. This is called Rabi-flopping. I mentioned before the discussion by P. C. W. Davies.[33] Davies's derivation shows clearly how increasing the number of degrees of freedom eliminates the interference terms and transforms the Rabi oscillation into an exponential decay. Similarly, the Daneri–Loinger–Prosperi proof that I described in Section 1 of this chapter relies on the large number of degrees of freedom of the measuring apparatus. This is the ground for their assumption (corresponding to assumption A_2 in 'Superposition and Macroscopic Observation') that there will be no correlation in time between systems originating in different microstates of the same macro-observable. This is exactly analogous to Haken's assumption, quoted in the last section, that b_λ^* and b_λ are uncorrelated over time, and it plays an analogous role. For that is just the assumption

[33] Davies, op. cit., in footnote 7.

that allows Haken to eliminate the interference terms for the photons and to derive the classical rate equations.

The most general proofs of this sort I know are in derivations of the quantum statistical master equation, an equation analogous to the evolution equations of classical statistical mechanics. The Markov treatment of radioactive decay from Essay 6 is a special case of this kind of derivation. In deriving the master equation the quantum system is coupled to a reservoir. In theory the two should evolve into a superposition in the composite space; but the Markov approximation removes the interference terms and decouples the systems. Again the Markov approximation, which treats the observation time as infinitely long, is justified by the large number of independent degrees of freedom in the reservoir, which gives rise to short correlation times there.

There are two difficulties with these sorts of proofs as a way of solving the characterization problem. The first is practical. It is a difficulty shared with the approach I shall defend in the end. Reduction of the wave packet, I have argued, takes place in a wide variety of circumstances. But treatments of the sort I have been describing have been developed for only a small number of cases, and, with respect to measurement in particular, treatments like that of Daneri, Loinger, and Prosperi are very abstract and diagrammatic. They do not treat any real measurement processes in detail.

The second difficulty is one in principle. Even if treatments like these can be extended to cover more cases, they do not in fact solve the characterization problem. That problem arises because we postulate two different kinds of evolution in nature and we look for a physical characteristic that determines when one occurs rather than the other. Unfortunately the characteristic we discover from these proofs holds only in models. It is not a characteristic of real situations. To eliminate the interference the number of relevant degrees of freedom must be *infinite*, or correlatively, the correlation time *zero*. In reality the number of degrees of freedom is always finite, and the correlation times are always positive.

Conceivably one could take the opposite tack, and urge that *all* real systems have infinitely many degrees of freedom.

That leaves no ground to distinguish the two kinds of evolution either. The fact that this view about real systems is intrinsically neither more nor less plausible than the first suggests that the concept *relevant number of degrees of freedom* is one that applies only in models and not in reality. If we are to apply it to reality I think we had best admit that real systems always have a finite number of degrees of freedom.

A real system may be very large—large enough to model it as having infinitely many degrees of freedom, or zero time correlations—but this does not solve the characterization problem. That problem requires one to separate two very different kinds of change, and size does not neatly divide the world in pieces. This is a familiar objection: if bigness matters, how big is big enough? Exactly when is a system big enough for nature to think it is infinite?

Sheer size cannot solve the characterization problem as I have laid it out. But I now think that what I have laid out is a psuedo-problem. The characterization problem is an artefact of the mathematics. There is no real problem because there are not two different kinds of evolution in quantum mechanics. There are evolutions that are correctly described by the Schroedinger equation, and there are evolutions that are correctly described by something like von Neumann's projection postulate. But these are not different kinds in any physically relevant sense. We think they are because of the way we write the theory.

I have come to see this by looking at theoretical frameworks recently developed in quantum statistical mechanics. (E. B. Davies's *Quantum Theory of Open Systems*[34] probably represents the best abstract formalization.) The point is simple and does not depend on details of the statistical theory. Von Neumann said there were two kinds of evolutions. He wrote down two very different looking equations. The framework he provides is not a convenient one for studying dissipative systems, like lasers or radiating atoms. As we have seen, the Schroedinger equation is not well able to handle these; neither is the simple projection postulate

[34] E. B. Davies, *Quantum Theory of Open Systems* (New York: Academic Press, 1976).

given by von Neumann. Quantum statistical mechanics has developed a more abstract formalism, better suited to these kinds of problems. This formalism writes down only one evolution equation; the Schroedinger equation and the projection postulate are both special cases of this single equation.

The evolution prescribed in the quantum statistical formalism is much like Schroedinger evolution, but with one central difference: the evolution operators for quantum statistical processes form a dynamical semi-group rather than a dynamical group. The essential difference between a group and a semi-group is that the semi-group lacks inverses and can thereby give rise to motions which are irreversible. For instance, the radiating atom will decay irreversibly instead of oscillating forever back and forth as it would in Rabi-flopping. Correlatively, the quantum statistical equation for evolution looks in abstract like a Schroedinger equation, except that the operation that governs it need not be represented by a unitary operator. A unitary operator is one whose adjoint equals its inverse. The effect of this is to preserve the lengths of vectors and the angles between them. There are other mathematical features associated with unitarity as well, but in the end they all serve to block reduction of the wave packet. So it is not surprising that the more general quantum statistical framework does not require a unitary operator for evolution.

From this new point of view there are not two kinds of evolution described by two kinds of equation. The new formalism writes down only one equation that governs every system. Is this too speedy a solution to the characterization problem? Perhaps there is only one equation, but are there not in reality two kinds of evolution? We may agree that Schroedinger's equation and reduction of the wave packet are special cases of the quantum statistical law: a Schroedinger-like equation appears when a unitary operator is employed; a reduction of the wave packet when there is a non-unitary operator. Does this not immediately show how to reconstruct the characterization problem? Some situations are described by a unitary operator and others by a non-unitary operator. What physical difference corresponds to this?

There is a simple, immediate reply, and I think it is the right reply, although we cannot accept it without considering some questions about determinism. The reply is that no *physical* difference need explain why a unitary operator is used in one situation and why a non-unitary one is required in another. Unitarity is a useful, perhaps fundamental, characteristic of operators. That does not mean that it represents a physical characteristic to be found in the real world. We are misled if we take all mathematics too seriously. Not every significant mathematical distinction marks a physical distinction in the things represented. Unitarity is a case in point.

'How', we may still ask, 'does nature know in a given situation whether to evolve the system in a unitary or a non-unitary way?' That is the wrong question. The right, but trite, question is only, 'How does nature know how to evolve the system?' Well, nature will evolve the system as the quantum statistical equation dictates. It will look at the forces and configurations and at the energies these give rise to, and will do what the equation requires when those energies obtain. Imagine that nature uses our methods of representation. It looks at the energies, writes down the operator that represents those energies, solves the quantum statistical equation, and finally produces the new state that the equation demands. Sometimes a unitary operator will be written down, sometimes not.

I reject the question, 'How does nature know in a given situation whether to use a unitary or non-unitary operator?' That question presupposes that nature first looks at the energies to see if the operator will be unitary or not, and then looks to see which particular operator of the designated kind must be written down. But there is no need for the middle step. The rules that assign operators to energies need not first choose the kind of operator before choosing the operator. They just choose the operator.

I write as if unitarity has no physical significance at all. It has only mathematical interest. Is this so? No, because unitarity is just the characteristic that precludes reduction of the wave packet; and, as we have seen, reduction of the wave packet is indeterministic, whereas Schroedinger evolution is deterministic. This, according to the quantum theory, is a genuine physical difference. (Deterministic motions are continuous; indeterministic are discontinuous.) I do not want to deny that there is this physical difference; but rather to deny that there must be some general fact about the energies that accounts for it. There need be no general characteristic true of situations in which the evolution is deterministic, and false when the evolution is indeterministic. The energies that obtain in a given situation determine how the system will evolve. Once the energies are fixed, that fixes whether the evolution is deterministic or indeterministic. No further physical facts are needed.

If this interpretation is adopted, determinism becomes for quantum mechanics a genuine but *fortuitous* physical characteristic. I call it fortuitous by analogy with Aristotle's meeting in the market place. In the *Physics*, Book II, Chapter 5, Aristotle imagines a chance encounter between two men. Each man goes to the market place for motives of his own. They meet there by accident. The meeting is fortuitous because the scheme of motives and capacities explains the presence of each man separately, but it does not explain their meeting. This does not mean that the meeting was not a genuine physical occurrence; nor that anything happened in the market place that could not be predicted from the explanatory factors at hand—in this case the motions and capacities of the individuals. It means only that meetings as meetings have no characteristic cause in the scheme of explanation at hand. This does not show that there is some mistake in the scheme, or that it is incomplete. We may be very interested in meetings, or for that matter, determinism, or the money cycle; but this does not guarantee that these features will have characteristic causes. A theory is not inadequate just because it does not find causes for all the things to which we attend. It can be faulted only if it fails to describe causes that nature supplies.

Professor Florence Leibowitz suggests another way to understand the place of determinism in the interpretation I propose. She says, 'Your claim about unitarity is in effect an assertion that indeterministic evolutions ought to be seen as "primitive" for quantum mechanics, in the sense that behaving according to the law of inertia is a primitive for post-scholastic mechanics.'[35] The conventional von Neumann-based view of evolution sees deterministic evolution as the natural situation, Leibowitz suggests. Indeterministic evolution is seen as a departure from what would naturally occur, and hence requires a cause—an interaction with a reservoir, perhaps, or with a conscious observer. But on my understanding of the quantum statistical formalism, indeterministic motions are natural too. They are not perturbations, and hence do not require causes. This is exactly analogous to what happens to inertial motion in the shift from Scholastic to Newtonian mechanics. For the Scholastic, continued motion in a straight line was a perturbation and a cause was required for it. In the Newtonian scheme the continued motion is natural or, as Leibowitz says, 'primitive'. Likewise quantum mechanics does not need a physical property to which unitarity corresponds: even if there were such a property, Leibowitz points out, 'it would not have any explanatory work to do'.

Throughout these essays I have urged that causality is a clue to what properties are real. Not all predicates that play a significant role in a theory pick out properties that are real for that theory. Many, for example, represent only properties in models, characteristics that allow the derivation of sound phenomenological laws, but which themselves play no role in the causal processes described by those laws. Unitarity is a different kind of example.

To understand what role unitarity plays in the theory, first look at another property that evolution operators may have: invariance under some group of transformations. In Schroedinger theory, the Hamiltonian describes the energies in the situation and thereby determines the evolution operator. Whenever the Hamiltonian is invariant under a group of transformations, there will be some constant of the motion which exhibits degeneracies; that is, different, incompatible states will have the same value for that quantity. Rotational invariance is a simple example. Rotational invariance in a Hamiltonian corresponds to spherical symmetry in the energies and forces represented by that Hamiltonian. The

[35] F. Leibowitz, private correspondence, 29 December 1981.

spherical symmetry produces degeneracies in the energy levels, and disturbing the symmetry eliminates the degeneracies. If a small non-symmetric field is added, a number of closely spaced lines will appear in the spectroscope where before there was only one. The rotational invariance of the Hamiltonian is a sign of a genuine physical characteristic of the situation.

Unitarity is different. We are often interested in whether the motion in a given kind of situation will be deterministic or indeterministic. There is a long way to find this out: consider the energies in the situation; write down the operator that represents them; solve the quantum statistical equation; and look to see if the consequent change is continuous. This is cumbersome. We would like some way to read from the operator itself whether the solutions will be continuous or not. Unitarity is a mark we can use, and it is one of the strengths of the particular mathematical structure we have that it gives us such a mark. Unitarity and rotational invariance are both significant features of the evolution operator, features that we single out for attention. But they play very different roles. Rotational invariance marks a genuine characteristic of the energy, a characteristic that is postulated as the source of a variety of physical consequences; unitarity provides an important mathematical convenience. To demand a physical correlate of unitarity is to misunderstand what functions it serves in the quantum theory.

I do not want to insist that unitarity does not represent a genuine property, but rather that the failure to find such a property is not a conceptual problem for the theory. Larry Laudan's description of a conceptual problem fits here. Laudan says, 'Such problems arise for a theory, T, . . . when T makes assumptions about the world that run counter to . . . prevailing metaphysical assumptions'.[36] Under von Neumann's proposal that quantum systems evolve under two distinct laws, some feature was required to signal which law should operate where. No physical characteristic could be found to serve, and the theory seemed driven to metaphysically

[36] L. Laudan, 'A Problem Solving Approach to Scientific Progress', in I. Hacking (ed.), *Scientific Revolutions* (Oxford: Oxford University Press, 1981) p. 146.

suspicious characteristics—fictional properties like infinite degrees of freedom or zero correlation times, or, even worse, interaction with conscious observers. But if the quantum statistical formalism can be made to work, no such property is required and the theory will not run counter to the 'prevailing metaphysical assumptions' that neither sheer size nor consciousness should matter to physics.

The *if* here presents an important condition. The comments of Jeremy Butterfield about my proposal seem to me right. Butterfield says,

Quantum statistical mechanics has provided a general theory of evolution of quantum states (pure and mixed) that encompasses [von Neumann's] two sorts of evolution, and many others, as special cases. Nor is this just a mathematical umbrella. It allows us to set up detailed models of phenomena that cannot be treated easily, if at all with the traditional formalism's two sorts of evolution.'[37]

Quantum optics is one place these detailed models have been developed, especially in the study of lasers. But we have seen that a broad range of cases make trouble for the traditional Schroedinger account—scattering, for example, or any situation in which pure states are prepared, and finally the issue that started us off—measurement. Much later Butterfield continues:

I do *not* want to pour cold water on this programme [of Cartwright's]; I find it very attractive. But I want to stress that it *is* a programme, not a *fait accompli*. To succeed with it, we need to provide detailed analyses of measurement situations, showing that the right mixtures are forthcoming. We need not of course cover all measurement situations, but we need to make it plausible that the right mixtures are generally obtained. (And here 'generally' need not mean 'universally'; it is the pervasiveness, not necessarily universality, of definite values that needs to be explained.) Only when we have such detailed analyses will the measurement problem be laid to rest.[38]

Butterfield gives good directions for what to do next.

I recommended the book of E. B. Davies as a good source for finding out more about the quantum statistical approach.

---

[37] J. Butterfield, 'Reply to N. Cartwright's 'How the Measurement Problem is an Artefact of the Mathematics', in R. Swinburne (ed.), *Space, Time, and Causality* (Dordrecht: D. Reidel, forthcoming).
[38] Ibid.

I should mention that Davies himself does not use the formalism in the way that I urge; for he is at pains to embed the non-unitary evolutions he studies into a Schroedinger evolution on a larger system. This goes along with his suggestion that non-unitarity is a mark of an *open* system, one which is in interaction with another. Open systems are presumably parts of larger closed systems, and these in Davies's account always undergo unitary change. I think this view is mistaken, for the reasons I have been urging throughout this essay. If the wave packet is not reduced on the larger system, it is not in fact reduced on the smaller either. The behaviour of the smaller system at best *looks* as if a reduction has occurred, and that is not good enough to account for measurements or for preparations.

I have been urging that, if the quantum statistical programme can work, the measurement problem becomes a psuedo-problem. But other, related, problems remain. These are the problems of how to pick the right operators, unitary or no, to represent a given physical situation. This is the piece by piece work of everyday physics, and it is good to have our philosophical attentions focused on it again. This is what on-going physics is about, and it knows no general procedure. In quantum mecahnics the correspondence principle tells us to work by analogy with classical mechanics, but the helpfulness of this suggestion soon runs out. We carry on by using our physical intuitions, analogies we see with other cases, specializations of more general considerations, and so forth. Sometimes we even choose the models we do because the functions we write down are ones we can solve. As Merzbacher remarks about the Schroedinger equation:

Quantum dynamics contains no general prescription for the construction of the operator H whose existence it asserts. The Hamiltonian operator must be found on the basis of experience, using the clues provided by the classical description, if one is available. Physical insight is required to make a judicious choice of the operators to be used in the description of the system . . . and to construct the Hamiltonian in terms of these variables.[39]

---

[39] Merzbacher, op. cit. footnote 11, p. 336–7.

As I argued in Essays 7 and 8, not bridge principles, but physical insights, are required to choose the right operators. But at least the quantum statistical programme offers hope that this mundane, though difficult, job of physics is all that there is to the measurement problem.

APPENDIX; AN EXPERIMENT TO TEST REDUCTION OF THE WAVE PACKET

In 1972 P. H. Eberhard considered non-unitary theories of the kind I endorse here and proposed four types of tests for them. I shall discuss in detail his tests involving the optical theorem of scattering theory, since this is the case I understand best, and it is a case that fits nicely with the discussion earlier in this essay. Eberhard tells us, about the theory he discusses, 'Our non-unitary theory resembles the description of quantic systems in interaction with a measurement apparatus, but no apparatus is involved in the physical processes that our theory is applied to'.[40] Eberhard calls theories that respect unitarity *class-A* theories. He will be concerned with a particular class of non-unitary theories— *class-B* theories. These are theories which model change in quantum systems on what happens in a complete measurement. Specifically, for an observable $M = \Sigma m \, |\phi_m\rangle\langle\phi_m|$, a B-type interaction takes the state D into D':

$$D \rightarrow D' = \sum_m |\phi_m\rangle\langle\phi_m| \, |D| \, |\phi_m\rangle\langle\phi_m|.$$

B-type theories are thus exactly the kind of theory I have urged, in which transitions genuinely occur into eigenstates $\{\phi_m\}$, and the final state for the ensemble is a classical mixture in which the ϕ_m do not interfere.

Eberhard tests type-B theories with the optical theorem. He tells us,

The optical theorem is derived from the principle that the wave scattered in the forward direction interferes with the incident wave in such a way as to conserve probabilities. If the forward scattering contained a mixture, i.e., non-interfering components, the test of the

[40] Eberhard, op. cit. footnote 31, p. 1.

optical theorem would fail. That test involves measurements of the differential cross section in the forward direction, including the interference region between Coulomb and strong interaction scatterings. The results can then be compared to the measurements of the total cross section.[41]

Eberhard looks at elastic scattering in a $\pi^- p$ interaction at 1.015, 1.527 and 2.004 GeV. The results agree with the prediction of the optical theorem 'within ±3%, when averaged over the three momenta'.[42] This agreement is good enough. If Eberhard's analysis is correct, class-*B* theories are ruled out for scattering interactions, and if they do not hold for scattering, they are not very plausible anywhere.

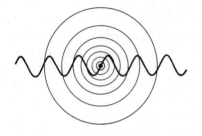

FIG 9.6. Scattering from a stationary target

The optical theorem is an obvious place to look for a test of non-unitary evolution. Consider a typical textbook presentation of the formal theory of scattering. I use Eugen Merzbacher's *Quantum Mechanics*. Merzbacher tells us 'The scattering matrix owes its central importance to the fact that it is *unitary*'[43] and later 'From the unitary property of the scattering matrix, we can derive an important theorem

[41] P. H. Eberhard, 'Tests of Unitarity', in A. Zichichi (ed.), *Progress in Scientific Culture. The Interdisciplinary Journal of the Ettore Majorana Centre. Winter 1976* (Trapani, Italy: Tipografia 'Cartograf', 1977).

[42] P. Eberhard; R. D. Tripp; Y. Declais; J. Seguinot; P. Baillon; C. Bricman; M. Ferro-Luzzi; J. M. Perrau; and T. Ypsilantis, 'A Test of the Optical Theorem', *Physics Letters* 53B, no. 1 (1974), p. 121.

[43] Merzbacher, op. cit. footnote 11, p. 501, italics in original.

for the scattering amplitudes'[44]—the optical theorem. Nevertheless the optical theorem does not rule out class-B theories. The optical theorem, I will argue, holds good in just the kind of class-B theory I have described for scattering.

In the case of elastic scattering, where the bombarding particle neither loses nor gains energy, the asymptotic state for large r for an outgoing particle whose initial momentum is $\mathbf{k}$, is given by

$$\psi'_{\mathbf{k}}(\mathbf{r}) \sim \frac{1}{2\pi^{3/2}} (\exp(i\mathbf{k} \cdot \mathbf{r}) + 1/r \exp(ikr) f_{\mathbf{k}}(\mathbf{r})). \qquad (1)$$

This state is a superposition of the original momentum eigenstate $\exp(i\mathbf{k}\cdot\mathbf{r})$ and an outgoing spherical wave $1/r\exp(ikr)$ as in Figure 9.6. The quantity $f_{\mathbf{k}}(\mathbf{r})$ is called the scattering amplitude. It is the imaginary part of the scattering amplitude in the forward direction—$Imf_k(0)$—which enters the optical theorem. As we can see from Figure 9.6, in the forward direction the original unscattered wave and the outgoing spherical wave interfere. The interference subtracts from the probability of the incoming wave in the forward direction. This is as we would expect, since the beam in the forward direction will be depleted by the particles that strike the target and are scattered.

It is easiest to calculate the interference if we switch to the formal theory of scattering. Equation (1) is the wave function version, for large r, of the Lippman-Schwinger equation:

$$\psi^+_k = \psi_k + \sum_n \frac{1}{E_s - H_0 + i\hbar\alpha} \psi_n T_{nk}.$$

Here the momentum states $\{\psi_n\}$ are eigenstates of the unperturbed Hamiltonian, H_0, and it is understood that the limit $\alpha \to 0$ is to be taken at the end of the calculation. The transition matrix T_{nk} is proportional to the scattering amplitude. We are interested in what proportion of the beam will be travelling in the forward direction after scattering, so we must calculate the probability $|\langle\psi_{\mathbf{k}}|\psi^+_{\mathbf{k}}\rangle|^2$. Substituting

[44] Ibid., p. 505.

$$T_{kk'} = \frac{2\pi\hbar^2}{\mu L^3} f_k(\hat{k}')$$

and using the fact that

$$\lim_{\alpha \to 0} \frac{1}{\omega + i\alpha} = \pi\delta(\omega)$$

and that

$$\delta(E_k - E_{k'}) = \frac{\mu}{\hbar^2 k} \delta(k - k'),$$

we get

$$|\langle \psi_k | \psi_k^+ \rangle|^2 = 1 + \frac{8\pi^4}{k^3 L^6} |f_k(0)|^2 + \frac{4\pi^3}{k L^3} Im\, f_k(0). \quad (2)$$

Now we may repeat the kind of classical argument that we considered for the two-slit experiment. A particle travelling in the forward direction after passing the target was either scattered from the target, or passed the target unscattered:

$$K = K\,\&\,(7S \vee S).$$

Since S and $7S$ are disjoint events

$$\text{Prob } K = \text{Prob}(K/7S)\,\text{Prob}(7S) + \text{Prob}(K/S)\,\text{Prob}(S). \quad (3)$$

But equation (2) shows that this classical reasoning will not do. The first two terms in equations (2) and (3) are identical, but, as in the two-slit experiment, the quantum mechanical calculation differs from the classical one by the interference terms, which are responsible in the two-slit case for the troughs in the diffraction pattern, and in scattering for the shadow cast by the target. We see that the amount of interference depends on the imaginary part of the forward scattering amplitude.

The optical theorem relates the total cross section, σ, to $Im\, f_k(0)$:

$$\text{Optical Theorem: } \sigma = \frac{4\pi}{k} Im\, f_k(0).$$

The cross section, σ, measures the total probability for scattering into any angle. Recall that Eberhard reported, 'The optical theorem is derived from the principle that the wave scattered in the forward direction interferes with the incident wave in such a way as to conserve probabilities'. We can now see why this is so. The optical theorem says that the loss in the beam in the forward direction, which we have seen comes from the interference term, $Im\ f_{\mathbf{k}}(0)$, is equal to the total number of particles scattered. Interference is thus an essential part of the optical theorem. How then can I maintain that reduction of the wave packet after scattering is consistent with the optical theorem?

The key to the answer is that one must be careful about what final states the system is supposed to enter when reduction occurs. I suggested, following Bohm, that after scattering each particle will be travelling in a specific direction, and with a specific energy. The reduction takes place into the eigenstates of momentum. The optical theorem precludes only a reduction into the pair scattered–unscattered. But the momentum probabilities already contain the interference between the incoming plane wave and the scattered spherical wave.

We can see this by looking back to the Lippman–Schwinger equation. It follows from that equation (taking the limit as $\alpha \to 0$) that the amplitude for a system with initial momentum $\mathbf{k}$ at $t = -\infty$ to have momentum $\mathbf{k}'$ at $t = +\infty$, is given by

$$S_{\mathbf{k}\mathbf{k}'} = \delta_{\mathbf{k}\mathbf{k}'} + \frac{4\pi^2 i}{kL^3}\ \delta(k - k')\, f_{\mathbf{k}}(\hat{\mathbf{k}}').$$

Here I have identified this amplitude with the $\mathbf{k}$, $\mathbf{k}'$th element of the *scattering matrix, S*, as we learn we can do from the formal theory of scattering. The total amplitude is thus a superposition of the amplitude from the scattered wave plus the amplitude from the unscattered wave, so the interference between the two is present right in the momentum amplitudes. When the complex conjugate is taken, equation (2) will result as required. It is no surprise then that the optical theorem still holds for the kind of reduction I propose.

Formally, I imagine that after reduction the state of particles in the beam is given by D':

$$D' = \sum_{\mathbf{k}'} |\psi_{\mathbf{k}'}\rangle\langle\psi_{\mathbf{k}'}| U(-\infty, +\infty)|\psi_{\mathbf{k}}\rangle\langle\psi_{\mathbf{k}}|$$

$$\times |U(-\infty, +\infty)|\psi_{\mathbf{k}'}\rangle\langle\psi_{\mathbf{k}'}|$$

$$= \sum_{\mathbf{k}'} S_{\mathbf{k}'\mathbf{k}} S_{\mathbf{k}'\mathbf{k}}^* |\psi_{\mathbf{k}'}\rangle\langle\psi_{\mathbf{k}'}|,$$

where $U(t, t')$ is the normal unitary evolution operator supplied by Schroedinger theory. (It is conventional to take the limits as $t \to \pm\infty$ since the times involved before detection are very long on the microscopic scale.) Since $D \to D'$ is a measurement-type interaction on the momentum eigenstates, the momentum probabilities after reduction will be the same as before reduction. But the optical theorem is a trivial consequence of the conservation of total probability among the momentum states. Here is where the unitarity of the scattering matrix, which Merzbacher stresses, plays its role. Because S is unitary, the probabilities to be in one or another of the momentum eigenstates sum to one

$$\sum_{\mathbf{k}'} S_{\mathbf{k}'\mathbf{k}} S_{\mathbf{k}'\mathbf{k}}^* = 1$$

not only in the unreduced state but in the reduced state as well. This is enough to guarantee that the optical theorem holds. The proof is simple, and I shall omit it here. (It is set as exercise 19.5 in Merzbacher: 'Derive the optical theorem from the conservation of probability and (19.12)',[45] where equation (19.12) gives the amplitude for the $\mathbf{k}'$th momentum state at t.) Thus, the optical theorem is consistent with a class-B interaction in which the wave packet is reduced into momentum eigenstates after the particle has been scattered.

What then of Eberhard's claims? To see how to reconcile what Eberhard says with the facts I have just noted, we need to look more closely at the kind of class-B theory which Eberhard considers. Eberhard notes that a non-unitary evolution like $D \to D'$ can always be written as a sum of unitary evolutions. This gives rise to something like a hidden

[45] Ibid., exercise 19.5, p. 507.

variable theory: where we see a single physical process which appears to follow a non-unitary rule, there are in fact a mixture of different processes each manifesting a unitary Schroedinger evolution. Or, alternatively: take the final pure states produced by Eberhard's set of 'component' unitary evolutions and wind them backwards by using the inverse of the original unitary scattering matrix. Then, the Eberhard style hidden variable theory says that, contrary to the normal assumption, the incoming state is not pure, but instead a mixture of these wound-backward states. Each state behaves exactly as the Schroedinger equation predicts. We end with a mixture, but only because we begin with one.

Even though he does not explicitly say so, Eberhard's calculations take this hidden variable theory very seriously. Eberhard's test uses two theorems from scattering theory. The first relates the differential cross-section in the forward direction to the scattering amplitude in that direction:

$$d\sigma(0) = |f_{\mathbf{k}}(0)|^2 \, d\Omega$$

Using R and J as Eberhard does to refer to the real and imaginary parts of $f_{\mathbf{k}}(0)$, we get Eberhard's equation (4.2).[46]

$$E(4.2): \quad d\sigma/d\Omega = R^2 + J^2.$$

The second theorem is the optical theorem, which Eberhard writes as

$$E(4.3): \quad J = k\sigma/4\pi\hbar.$$

As Eberhard remarks, R can be calculated from J, or it can be determined from the interference with Coulomb scattering. Since σ and $d\sigma(0)/d\Omega$ can be measured in independent experiments, a test of the optical theorem is at hand.

Let us now look to see how Eberhard turns this into a test of class-B evolutions using his earlier theorem that any non-unitary evolution of B-type is equivalent to a weighted average of unitary changes. Eberhard claims:

In a class B theory, there are pseudo-states j that correspond to weights w_j and unitary matrices S_j. Each unitary matrix S_j corresponds to a class A theory, therefore to a σ_j, to a $d\sigma_j/d\Omega$, to a $R_j(E)$ and to a $J_j(E)$

[46] Eberhard, op. cit., note 30, p. 14.

satisfying eq. (4.1) to (4.5). The effective probability distributions are the weighted averages of those class A predictions and the effective cross sections σ and $d\sigma/d\Omega$ are the weighted averages of the σ_j's and of the $d\sigma_j/d\Omega$ respectively.[47]

So, says Eberhard,

$$E(4.7): \quad d\sigma(0)/d\Omega = \sum_j w_j \, d\sigma_j/d\Omega$$

$$= \sum_j w_j \, R_j^2 + J_j^2$$

and

$$E(4.8)a: \quad J = \sum_j w_j J_j = \frac{k}{4\pi} \sum_j w_j \sigma_j = \frac{k\sigma}{4\pi\hbar}$$

$$E(4.8)b: \quad R = \sum_j \left(w_j R_j \right)$$

Using $E(4.8)$, $R^2 + J^2 = (\Sigma \, w_j R_j)^2 + (\Sigma \, w_j J_j)^2$. But in general

$$E.I.: \quad \left(\sum_j w_j R_j^2 + \sum_j w_j J_j^2 \right) \neq \left\{ \left(\sum_j w_j R_j \right)^2 + \left(\sum_j w_j J_j \right)^2 \right\}$$

So, if $E(4.7)$ is correct, $E(4.2)$ will be violated.

It remains to substantiate that equations $E(4.7)$ and $E(4.8)$ are true for class-B theories. $E(4.7)$ is straightforward. Letting W represent the total transition rate into a solid angle $d\Omega$, from B

$$W = \sum_{k' \in d\Omega} \sum_{k''} \frac{d}{dt} \langle \psi_{k'} | \psi_{k''} \rangle \langle \psi_{k''} | U | \psi_k \rangle \langle \psi_k | U | \psi_{k''} \rangle \langle \psi_{k''} | \psi_{k'} \rangle$$

$$= \sum_{k' \in d\Omega} \frac{d}{dt} \langle \psi_{k'} | U | \psi_k \rangle \langle \psi_k | U | \psi_{k'} \rangle$$

$$= \lim_{\Delta t \to 0} \frac{1}{\Delta t} \left(\sum_{k' \in d\Omega} \langle \psi_{k'} | U(t + \Delta t) | \psi_k \rangle \langle \psi_k | U(t + \Delta t) | \psi_{k'} \rangle \right.$$

$$\left. - \sum_{k' \in d\Omega} \langle \psi_{k'} | U(t) | \psi_k \rangle \langle \psi_k | U(t) | \psi_{k'} \rangle \right)$$

[47] Ibid., p. 14.

From Eberhard's earlier theorem that class-B evolutions are equivalent to weighted averages of unitary evolution U_j,

$$W = \lim_{\Delta t \to 0} \frac{1}{\Delta t} \times$$

$$\times \left[\sum_{k'} \left(\sum_j w_j \langle \psi_{k'} | U_j(t + \Delta t) | \psi_k \rangle \langle \psi_k | U_j(t + \Delta t) | \psi_{k'} \rangle \right. \right.$$

$$\left. \left. - \sum_j w_j \langle \psi_{k'} | U_j(t) | \psi_k \rangle \langle \psi_k | U_j(t) | \psi_{k'} \rangle \right) \right]$$

$$= \sum_j w_j \lim_{\Delta t \to 0} \frac{1}{\Delta t} \sum_{k'} \langle \psi_{k'} | U_j(t + \Delta t) | \psi_k \rangle \langle \psi_k | U_j(t + \Delta t) | \psi_{k'} \rangle$$

$$- \langle \psi_{k'} | U_j(t) | \psi_k \rangle \langle \psi_k | U_j(t) | \psi_{k'} \rangle$$

$$= \sum_j w_j W_j.$$

But

$$d\sigma \underset{\mathrm{d\,f}}{=} W/(\hbar k/\mu L^3),$$

where $\hbar k/\mu L^3$ is the probability that a particle is incident on a unit area perpendicular to the beam per unit time. Hence,

$$d\sigma = \frac{\mu L^3}{\hbar k} \sum w_j W_j = \sum w_j d\sigma_j.$$

So equation E(4.7) holds.

But what of E(4.8)? E(4.8) is justified only if we insist on an Eberhard-style hidden variable theory. Eberhard shows that the non-unitary evolution B can always be mathematically decomposed as an average over unitary evolutions. The hidden variable version of the class-B theory supposes that this mathematical decomposition corresponds to a physical reality: scattering is really a mix of physical processes, each process governed by one of the unitary operators, U_j, which make up the mathematical decomposition. In this case, we have a variety of different scattering processes, each with its own scattering amplitude, $f_k^j(k')$, and its own cross-sections, σ_j, and E(4.8) is a reasonable constraint on the real and imaginary parts of the scattering amplitudes.

But is this physically robust hidden variable version of a class-*B* theory a reasonable one from our point of view? No, it is not. For it does not solve the problem of preparation that motivated our class-*B* theory to begin with. Scattering interactions prepare beams in momentum eigenstates: particles which appear in a particular solid angle dΩ at one time are expected—those very same particles—to be travelling in exactly the same solid angle later, unless they are interfered with. So we look for physical processes whose end-states are eigenstates of momentum. But the end-states of the processes U_j are nothing like that.

From Eberhard's decomposition proof there are as many 'hidden' processes as there are dimensions to the space needed to treat the system. In the case of scattering, a 'quasi-continuum' of states is required. Each 'hidden process' turns out to be itself a scattering-type interaction. The end-state of the first hidden process is just the normal outgoing spherical wave of conventional scattering theory. This conventional state will have the weight $1/n$, when n is the dimension of the space. So, too, will each of the remaining states. The end-state of the second process will be much like the first, except that the amplitude of the second momentum eigenstate will be rotated by 180°. Similarly, the third process rotates the amplitude of the third momentum eigenstate by 180°; the fourth rotates the fourth amplitude; and so on. In average, the effect of these rotations is to cancel the interference among the momentum states and to produce a final mixture whose statistical predictions are just like those from a mixture of momentum states. But in truth, when the hidden variable account is taken seriously, the final state is a mixture of almost spherical waves, each itself a superposition of momentum eigenstates; and it is not in fact a mixture of momentum eigenstates as we had hoped. But if we do not take the hidden variable theory seriously and give a physical interpretation to the decomposition into unitary processes, there is no ground for equations E(4.8)a and E(4.8)b, and the optical theorem is no test of *B*-type evolutions.

The Eberhard inequality E.I. is based on equations E(4.8)a and E(4.8)b, which I claim are plausible for an Eberhard–style hidden variable theory, but which do not hold for a class-B theory which takes each incoming particle into a momentum eigenstate. We should now confirm this last claim. The process which produces a mixture of momentum eigenstates is itself composed of a mixture of processes, each of which produces one or another of the eigenstates of momentum as its final product. (Note: each of these processes is *non*-unitary, because it shrinks the vector. Nor can we reconstruct it as a unitary evolution by taking the shrinking factor, as Eberhard does, as a weight with which a unitary change into a momentum eigenstate might occur, because the 'weights' for the various processes would depend not on the nature of the interaction but on the structure of the incoming state.) We need to be sure that E.I. does not follow equally on my account as on the hidden scattering account. But this is easy. The 'weights' here are each 1. Each interaction scatters entirely into a single direction, and it is entirely responsible for all the scattering that occurs into that direction. So $f_{\mathbf{k}}^j(\mathbf{k}') = f_{\mathbf{k}}(\mathbf{k}')\,\delta_{j\mathbf{k}'}$ and hence

$$\left(\sum w_j\, f_{\mathbf{k}}^j(0)\right)^2 = \left(\sum 1\, f_{\mathbf{k}}^j(0)\delta_{j\mathbf{k}}\right)^2 = |f_{\mathbf{k}}(0)|^2$$

$$= \sum 1\,(f_{\mathbf{k}}(0)\delta_{j\mathbf{k}})^2 = \sum w_j(f_{\mathbf{k}}^j(0))^2.$$

So E.I. does not plague the view that combines reductions into different momentum eigenstates to get the mixture D', though it does plague the composition of hidden scattering processes, as Eberhard shows. But this latter view is not one I would incline to, since it is not one that allows for the preparation of momentum eigenstates in scattering interactions.

Author Index

Agarwal, G. S. 79, 123, 124, 125
Airy, G. B. 2
Alexander, P. 55
Aristotle 9, 110, 201
Armstrong, T. A. 130

Boyle, R. 100
Bell, J. S. 168
Becquerel, H. 192
Bethe, H. 122, 123
Bickel, P. 37
Blalock, H. 23
Bohm, D. 191, 192
Bohr, N. 193, 194
Born, M. 180, 191
Bromberger, S. 75
Butt, D. K. 118
Butterfield, J. 204

Carnap, R. 26
Cohen, M. 24
Creary, L. 62, 63, 65, 66, 67
Crookes, W. 5, 81
Curie, M. 192
Curie, P. 192
Cushing, J. 7, 158

Daneri, A. 169, 170, 295, 196, 197
Davies, E. B. 198, 204, 205
Davies, P. C. W. 169, 196
Descartes, R. 75
Donagan, A. 16
Duhem, P. 4, 19, 76, 77, 87, 88, 89,
    90, 91, 92, 93, 94, 96, 97, 159

Eberhard, P. H. 195, 206, 207, 210,
    211, 212, 213, 214, 215, 216
Einstein, A. 21, 193
Ennis, R. 77
Everitt, C. W. F. 2, 5, 6, 82, 148

Faraday, M. 2
Fermi, E. 151
Feynman, R. 19, 55, 56, 57, 59, 161,
    162, 166

Gardner, M. 177

Gibbard, A. 33
Gibbins, P. 177
Goldberger, M. 117
Goldman, M. 81
Grice, H. P. 33, 128, 129, 153
Grünbaum, A. 94, 101, 102, 131

Hacking, I. 2, 20, 88, 98
Haken, H. 79, 80, 190, 196, 197
Hamilton, W. 135
Hammel, E. 37
Hanson, B. 45
Harman, G. 75, 85
Harper, W. 38
Hausman, D. 77
Hawking, S. 12
Hempel, C. G. 26, 27, 28, 29, 30, 44,
    45, 94, 101, 131, 132, 134, 155
Hesse, M. 157
Hooke, R. 100
Hume, D. 13, 61, 74

Infeld, L. 21

Jackson, F. 40
Jaynes, E. T. 118
Jeffrey, R. 27, 34, 103
Joseph, G. 12

Kant, I. 88
Klauder, J. 148, 149, 158
Klein, M. 46, 47
Kline, S. 63
Kochen, S. 168
Kruskal, S. 168
Kuhn, T. 143

Lamb, W. 116, 122, 123, 125, 186,
    188, 189
Laudan, L. 94, 97, 203
Leibowitz, F. 201, 202
Loinger, A. 169, 170, 195, 196, 197
Lorentz, H. 2
Louisell, W. 78, 115, 146, 147, 148,
    150, 151, 189

Mackey, G. W. 135

Mapleton, R. 121
Margenau, H. 182
Maxwell, J. 2, 5, 6, 8, 9, 11, 12, 154, 155, 156
Merzbacher, E. 113, 136, 141, 142, 174, 175, 178, 207, 211
Messiah, A. 67, 68, 136, 137, 138
Mill, J. S. 7, 58, 60, 61, 70

Nagel, E. 24, 101, 131
Newton, I. 75
Nordby, J. 13, 14, 15, 103, 107, 111
Norman, S. 71

O'Connell, J. W. 37
Onsager, L. 65

Pais, A. 118
Pauling, L. 166
Pearson, K. 39
Perrin, J. 82, 83, 84
Perry, J. 54
Piccolomini 90, 91
Prosperi, G. M. 169, 170, 195, 196, 197
Ptolemy, 90
Putnam, H. 56, 57, 77, 88, 110, 170, 177

Redhead, M. 158
Retherford, R. C. 116, 122, 125
Russell, B. 13, 21, 74, 112
Rutherford, E. 192, 193

Salmon, W. 16, 23, 24, 25, 26, 27, 28, 29, 30, 44, 52, 82, 103
Sargent, M. 186, 188
Schroedinger, E. 186
Scriven, M. 16, 75

Scully, M. 186, 188
Sellars, W. 157, 159, 160
Shiva, V. 195
Siegman, A. 129, 147, 148, 149, 187, 188
Simon, H. 23
Simpson, E. H. 7, 10
Skyrms, B. 43
Smart, J. J. C. 54
Smith, A. W. 130
Sneed, J. 159
Soddy, F. 192, 193
Specker, E. P. 168
Stairs, A. 181
Sudarshan, E. C. G. 148, 149, 158
Suppe, F. 131
Suppes, P. 23, 28, 29, 44, 52, 159

Thom, R. 75
Thomson, J. J. 61
Thucydides 139
Truesdell, C. A. 65

Van der Pol, B. 130
Van Dyke, M. 14
Van Fraassen, B. 4, 6, 8, 10, 56, 87, 88, 89, 90, 91, 92, 93, 94, 96, 97, 99, 159
Von Neumann, J. 163, 167, 194, 195, 198, 203

Watson, K. 117
Weisskopf, V. 113, 119, 122, 163
Wilson, A. R. 118
Winter, R. 118
Wollheim, R. 91

Yule, G. 24

Subject Index

Airy's Law 2
amplifiers 103, 104, **107-13**
approximations 3, 13 ff., 104 ff.,
 107-19, 110, 112, 113, 117,
 113-127, 133, 134
 ab vero 14
 ad verum 14, 15
'as if' operator 129, 130, 131, 133
Avogadro's hypothesis 84 ff.

Boltzmann's equation 6, 9, 11, 154
bridge laws (bridge principles) 12, 15,
 131 ff., **135-9**, 143, **144-51**,
 155, 156

causal behaviour (lasers) 147, 150
causal explanation 4, 5, 8, 10 ff., 15,
 27, 29, 36, **75-8**, 89, 93
 causal account 6, 82, 97, 98, 162
 causal role 8, 182, 191
 causal story 5, 9, 11, 12, 76, 77,
 79, 81, 82, 86, 161, 162
causal models 23
causal powers 59
causal principles (causal laws) 10,
 21-44, 72
causal process 4, 16, 19, 61, 84, 85,
 128, 152, 160, 162, 202
ceteris paribus laws (*ceteris paribus*
 principles) 3, 45, **46-7**, 57, 58,
 64, 155
characterization problem 169 ff., 199
classical electron oscillator 129, 130,
 148, 186
cloud chamber 92, 93, 99, 182
coincidence, argument from 75, **82-5**
composition (combination) of causes
 3, **11-15**, 51, **56-9**, 62, 66,
 69-72, 73
 see also interactions, causal *and*
 cross-effects
Coulomb force (Coulomb law) 57,
 59 ff., 66, 68 ff., 138, 159, 212
counterfactuals 34, 35, **38-40**, 69
covering law model of explanation
 11, 15 ff., 44, 45, 48 ff., 52, 69,
 155, 162

cross-effects 64, 65, 112
 see also composition of causes *and*
 interactions, causal

damping 77, **78-82**, 113, 149
decay, exponential 27, 93, **113-18**,
 122, 164, 169, 170, 179, 189,
 190, 192 ff., 196, 197, 199
decision theory **33-40**
deductive-nomological model of explanation 17, 44, 94 ff., 101 ff.,
 110, 112, 127, 151, 161, 162
degrees of freedom 150, 196 ff., 204
determinism (indeterminism) 21, 52,
 164, 167, 179, 191, 192, 200 ff.
diffraction 3, 172, 175
dipole radiation 186 ff.

effectiveness 33 ff., 42, 43
eigenvalue-eigenvector link 168
Einstein-Podolsky-Rosen paradox 18
experiments 6, 7, 20, 82 ff., 98, 160,
 161
experimental test of exponential law
 118
explanatory power 10, 20, 34, **56-9**,
 72, 139, 144, 152
exponential decay, *see* decay, exponential

facticity 54, 58, 61, 62, 65, 71 ff., 152
Fick's law 63, 64
Fokker-Planck equation 130
forces, component and resultant 59,
 60, 66, 69, 70
fortuitous characteristic 201

generic-specific account 94, 103, 104,
 106, 107, 110, 126
gravitational force (gravitational law)
 56, 57, 59 ff., 66, 69 ff., 74

Hamiltonian, model 135, 136, 139,
 144, 159
hidden variables 168, 212, 214, 215
homogeneity, causal 25, 26, 28, 40, 41
Hume world **40-2**

idealization 48, 109 ff., 134, 136, 147, 148, 153, 155, 158
inductive–statistical model of explanation 26, 30, 44, 47
inference to best explanation 4, 6, 15, 75, 82, 83, 85, **87–9**
inference to most likely cause 6, 85, 92, 94
infinite potentials 150, 153
interactions, causal **30–2**, 63
    see also composition of causes and cross-effects
internal principles 131, 132, 135, 139

joint effects 43
    see also spurious cause
joint probability distribution 156, 177, 178, 181

Lamb shift 116, 117, 137, 138
    in the excited state **119–23**
    in the ground state 119, **123–7**
lasers 3, 79, 80, 98, 128 ff., 145, 146, 148, 149, 152, 153, 157 ff., 161, 186, 189, 198, 204
laws of association 10, 21, 25, 26, 28, 35, 40 ff., 46
line broadening 77 ff.

macroscopic observable (macroscopic system) 167, 168, 170 ff., 195, 196
Markov approximation 113 ff., 123, 124, 133, 146, 189, 197
master equation 79, 81, 113, 114, 123, 125, 145, 197
measurement problem 18, **162–5**
mechanical philosophy 100, 107
Mill's methods 7, 98
mixtures 168, 170, 171, 194, 204, 206, 212, 215
models, explanatory 4, 11, 12, 15, 17, 44, 83 ff., 95, 104, 107, 111, 129, 140, 143, 144, 148, 152, 154, 156 ff., 197, 198, 202

Newton's laws 63

observable (unobservable) 1, 2, 5, 6, 19, 56, 83, 100, 106, 159, 160
Ohm's law 63
old quantum theory (new quantum theory) 137, 192 ff.

open systems 198, 205
optical theorem 206 ff.
order of integration 120 ff.

partial conditional probability 42
partitioning 36 ff., 41, 43
Pauli equation 114, 115, 124
phenomenological laws 1 ff., 8, 11, 16, 19, 85, 88 ff., 93 ff., 100 ff., 106, 160, 161
phenomenological models (phenomenological terms) 148, 151
photographic plate 182 ff.
positions of quantum systems 163, 166, 167, 171, 178, 182, 183, 186, 188 ff.
position probabilities 164, 174, 179, 180, 188, 190, 191
preparation process (problem of preparation) 174, 194, 205, 215, 216
prepared descriptions (unprepared descriptions) 15, 17, 133, 139, 147, 160, 162
probability distributions, classical 153 ff.
probabilistic model of causation 44, 47
projection postulate 162, 167, 170, 198
property of convenience 153, 156

quantum electrodynamics (quantum field theory) 7, 8
quantum logic 164, 177, 181
quantum statistical mechanics 164, 199 ff., 204, 205

Rabi-flopping 196, 199
radioactivity 25, 27, 28, 78, 93, 118, 192, 193
    see also decay, exponential
radiometer 5, 6, 8, 9, 11, 81, 82, 154 ff.
rate equations 189, 190
realistic in the first sense (in the second sense) 150, 152
reduction of the wave packet 163, 164, 167, 170, 174, 194 ff., **206–16**, appendix
redundancy 76, 79, 90
reference class problem 28, 29
reservoir 113 ff., 129, 133, 146, 148 ff., 189, 197, 202

rotating wave approximation 119, 123 ff.

scattering 164, 179, 180, 182, 189, 194, 204, 206 ff.
Schroedinger evolution 164, 196, 200, 205, 212
semantical view of theories 159 ff.
Simpson's paradox 10, 24, 36 ff.
simulacrum account of explanation 4, 17, 48, **151–62**, 190
SLAC 172, 173
Snell's law 46 ff.
spurious cause 31, 33, 34
    see also joint effects
statistical relevance 30, 40, 44, 47
strategies 10, 21, 22, 33 ff., **36–8**, 43
super laws 12, 70, 71

test situation 26 ff., 30, 35

theoretical entities 6, 7, 56, **31–4**, 97, 99
theory-observation distinction 132, 133
transition process (transition probability) 164, **174–95**, 206
true probability **38–40**
two-slit experiment 175, 186, 195, 209

unitarity 18, 199 ff., 205 ff., appendix
unity of science (unity of nature) 12, 13, 95, 96

Van der Pol's equation 148, 153, 157
vector addition 59, 60, 62, 63, 69

Weisskopf–Wigner treatment 99, 113, 115, 117, 119, 122, 123, 196

zero-time correlations 150, 153, 196